Lectures on Natural Philosophy

MARGARET BRYAN

CAMBRIDGE
UNIVERSITY PRESS

CAMBRIDGE UNIVERSITY PRESS

Cambridge, New York, Melbourne, Madrid, Cape Town,
Singapore, São Paolo, Delhi, Tokyo, Mexico City

Published in the United States of America by Cambridge University Press, New York

www.cambridge.org
Information on this title: www.cambridge.org/9781108038089

This edition first published 1806
This digitally printed version 2011

ISBN 978-1-108-03808-9 Paperback

CAMBRIDGE LIBRARY COLLECTION

Books of enduring scholarly value

Women's Writing

The later twentieth century saw a huge wave of academic interest in women's writing, which led to the rediscovery of neglected works from a wide range of genres, periods and languages. Many books that were immensely popular and influential in their own day are now studied again, both for their own sake and for what they reveal about the social, political and cultural conditions of their time. A pioneering resource in this area is Orlando: Women's Writing in the British Isles from the Beginnings to the Present (http://orlando.cambridge.org), which provides entries on authors' lives and writing careers, contextual material, timelines, sets of internal links, and bibliographies. Its editors have made a major contribution to the selection of the works reissued in this series within the Cambridge Library Collection, which focuses on non-fiction publications by women on a wide range of subjects from astronomy to biography, music to political economy, and education to prison reform.

Lectures on Natural Philosophy

Margaret Bryan (c.1760–1816) taught natural science to women at a time when it was largely the preserve of men. She ran a boarding school for girls in Blackheath, London, from 1795 to 1806, and the curriculum included mathematics and sciences – rarely offered to young women. She published her lecture notes on astronomy in 1797, and after their positive reception she decided to undertake another volume of lectures. This resulting work, published in 1806, is a collection of Bryan's lectures on 'natural philosophy', containing thirteen chapters on topics such as mechanics, pneumatics and acoustics, magnetism and electricity. Each chapter provides illustrations, and at the end of the volume there is an appendix with astronomical and geographical questions and exercises, as well as a scientific glossary. These lectures provide a glimpse into the little-known world of women's education towards the end of the Georgian period.

Cambridge University Press has long been a pioneer in the reissuing of out-of-print titles from its own backlist, producing digital reprints of books that are still sought after by scholars and students but could not be reprinted economically using traditional technology. The Cambridge Library Collection extends this activity to a wider range of books which are still of importance to researchers and professionals, either for the source material they contain, or as landmarks in the history of their academic discipline.

Drawing from the world-renowned collections in the Cambridge University Library, and guided by the advice of experts in each subject area, Cambridge University Press is using state-of-the-art scanning machines in its own Printing House to capture the content of each book selected for inclusion. The files are processed to give a consistently clear, crisp image, and the books finished to the high quality standard for which the Press is recognised around the world. The latest print-on-demand technology ensures that the books will remain available indefinitely, and that orders for single or multiple copies can quickly be supplied.

The Cambridge Library Collection will bring back to life books of enduring scholarly value (including out-of-copyright works originally issued by other publishers) across a wide range of disciplines in the humanities and social sciences and in science and technology.

T. Kearsley, Pinxit.   Heath Sculpsit.

MARGARET BRYAN

Published May 10th 1806, by G. Kearsley, Fleet Street

DEDICATED, BY PERMISSION,

TO HER ROYAL HIGHNESS

THE PRINCESS CHARLOTTE OF WALES.

LECTURES

ON

NATURAL PHILOSOPHY:

THE RESULT OF MANY YEARS' PRACTICAL EXPERIENCE
OF THE FACTS ELUCIDATED.

WITH

AN APPENDIX:

CONTAINING;

A GREAT NUMBER AND VARIETY OF

ASTRONOMICAL AND GEOGRAPHICAL PROBLEMS:

ALSO

SOME USEFUL TABLES,

AND

A COMPREHENSIVE VOCABULARY.

———◆———

God speaks through all and is in all things found.

ROWE.

———◆———

BY MARGARET BRYAN.

LONDON:

Printed by Thomas Davison, Whitefriars;

AND SOLD FOR THE AUTHORESS BY GEORGE KEARSLEY, FLEET-STREET;
AND JAMES CARPENTER, OLD BOND-STREET.

———◆———

1806.

UNCONSCIOUS even of the possibility of the distinguishing honour now conferred on my mental exertions, by the truly gratifying and most exalted distinction they could receive—that of addressing these Lectures to Your Royal Highness—in no part of this work could any personal application of the instruction it is intended to convey be made to a Personage of such high importance; presumptuous, indeed, would have been such an appeal from any individual not appointed to the high and important office of forming the character of a Princess, already surrounded by the ablest instructors, and guided by their eminent wisdom and zeal.

That I am deemed worthy of presenting a work of science to Your Royal Highness, is a circumstance in the highest degree flattering to my abilities—for such a distinction would not have been permitted on slight pretensions.

With extreme satisfaction I offer these Lectures to Your Royal Highness's perusal; for the truths they inculcate, though of importance to all human nature, are peculiarly adapted to noble minds and persons of exalted stations; producing in these the most extensive and important effects.

May the blossoming genius of Your Royal Highness's mind ripen into maturity with that vigour and sweetness, as may render it a public blessing and delight, adorning Your elevated station, and diffusing a salutary and benign influence through Your native land!

Impressed with sentiments the most respectful and zealous, I beg leave to subscribe myself,

Your Royal Highness's

Most faithful, devoted

And obedient humble servant,

MARGARET BRYAN.

ADDRESS TO MY PUPILS.

NO conduct, in my opinion, is more unnatural and culpable, than that of disregarding the present and future interests and happiness of young people placed immediately under our care and guidance. The period of scholastic improvement, however protracted, must terminate; when very often the early lessons of a preceptress are too faintly impressed on the minds of youth to withstand the obliterating effects of time; and the friction of that polisher, which renders the surface smooth, but destroys the native, simple and charming characters of truth.

To prevent this great and fatal evil, I have provided you, my dear Pupils—by the evidences of the nature and attributes of the Deity pourtrayed in these Lectures—with a defence against the vain sophistry of the world; arming you with a perpetual talisman, which, your conduct justifying its power, will secure you from all pernicious doctrines, and guard your religious and moral principles against all innovations.

Let the expectations I have encouraged, my dear friends, be realized, for my heart is deeply interested in the result of my labours. Embrace my precepts, and resolutely endeavour to secure their solid advantages. Influenced by principle, I first wrote these Lectures; actuated by the same feeling, I now perpetuate them. Do not disappoint the hopes I have cherished, by negligence and inattention. Omit no opportunity of applying the lessons they impart to perfect every thing that gives either value or grace to the female character. Consider them, not merely as infusions of science, but also, and more particularly, as leading to all that is lovely, dignifying and noble. Cull the sweets of religion, as you rove through the flowery paths of Natural Philosophy; and endeavour so to imbibe and employ their salutary influence, as may afford the happy promise of finally obtaining those gratifications which are indescribable in their nature, perpetual in their duration. Let my tender solicitude be repaid with knowing, by the evidences displayed in your conduct through life, that I have been instrumental in fixing your character; then will all my anxiety be compensated, and I shall continue to prosecute with energy my task of duty, assured of my reward in your present and future felicity. Happy indeed am I, in the good effects my precepts have produced on many whom I have had the pleasure of instructing : and on none are these impressions more distinguishable than on my own dear children; in whose strict piety, candid integrity and dutiful affection, I rejoice most exceedingly, and feel more blessed than language can express. That the Almighty may continue to direct my dear children and pupils by his spirit—establish them in the performance of their duty by his aid—and endow them with constancy in bearing the afflictions of human life—is my most ardent supplication !

b

Impressed with this hope, and animated by the tenderest affection—I rejoice in the titles of Parent and Preceptress.

M. BRYAN.

PREFACE.

INTELLECTUAL acquirements are the support, the defence, and the glory of human nature: without a proper exercise of our reasoning faculties, we become a disgrace to society, a burthen to ourselves, and liable to all those deviations from rectitude which must finally doom us to obscurity, or involve us in disappointment. The most important knowledge in all situations of life, is that of our duty to God and Man; which the Scriptures teach, and all the attributes of the Deity most impressively enforce. Hence the truths of religion, and those of natural philosophy, by their affinity, combine in their effects, and unitedly enliven our present enjoyments, and strengthen and confirm our future hope. Actuated by these convictions, and encouraged by the very honourable and distinguishing patronage my Lectures on Astronomy experienced, I present this work on Natural Philosophy; which was originally written, and is principally intended, for the instruction and improvement of my pupils in the knowledge of physics: to impress them with a just sense of the attributes of the Deity; and thus, by un-equivocal deductions, to strengthen and confirm their moral and religious principles.

In the subjects relating to Astronomy I have been brief, through delicacy to the Proprietors of my former Lectures; wishing that this publication may not injure, but rather promote their interest. Yet, duly impressed with my duty to the Public, I have neglected nothing in this work that is essential to a clear and perfect elucidation of the phenomena to be explained by the subjoined Problems *.

That the knowledge imparted by the operations performed by means of the globes and maps may not be merely mechanical, but really scientific, I have associated the theoretical with the practical illustra-tions, and have adverted to diagrams; which rational and referential mode of elucidation will, I trust, render my communications more useful, and better under-stood, than the generality of similar performances.

So many learned men having communicated their

* For a more general illustration of celestial phenomena I refer my readers to my Lectures on Astronomy, octavo edition.

philosophical researches, and illustrated them by the sublimest mathematical demonstrations, it may appear vain and presumptuous in me to expect public notice: yet I trust the candid mind will not impute to me either vanity or arrogance; as I certainly offer myself merely as a reflector of the intrinsic light of superior genius and erudition; to moderate its effects in enquiring minds, not fully prepared to imbibe and sustain its profound mathematical energies.

After eight years' study of the facts I have attempted to investigate, aided by seven years' practical experience to establish these principles; I am emboldened to venture my claim to public attention and confidence. These Lectures have been revised under the disadvantages of a desultory mode of conducting that important business; for which the various professional claims on my time can alone apologize. These I beg may be accepted as an excuse for any imperfections in arrangement that may appear to the critical examiners of this work. For any deficiency of elegance in the diction, I must plead, that descriptions of apparatus and experiments do not admit of that graceful and decorated style of expression, which the sublime nature of the subjects, unfettered by scientific elucidation,

naturally demand : but this apparent disadvantage
in philosophical writing is more than compensated by
the simplicity and importance of the truths it explains;
which need no aid of ornament—being such as appeal
to the reason, not to the passions, to effect their essential and valuable purposes.

I trust I may venture to assert the veracity of the
scientific principles contained in these Lectures, without impeachment of my judgment and integrity ; as
they have received the sanction of my very learned
friend and patron, whose approbation of my ideas could
alone induce me to give them currency. For though
I may be able to appreciate the qualities of natural
objects, yet I could never assume sufficient confidence
in my own powers and capacities to assert my claim to
public attention.

Through the whole of the anatomical parts of these
Lectures I have followed the very excellent divine Dr.
Paley, in his Natural Theology :—a work comprehensive
in its nature, important in its application, and extensive
in its elucidations of the divine wisdom and omnipotence of our great Creator ; tending to the subversion
of all irreligious principles and doctrines, by the most

substantial evidences of the divine origin of religion; and inviting us to that consideration of our nature and our dependence on Divine power and goodness, which cannot fail to confirm us in the true belief and service of God.

I now with less timidity than formerly submit myself to the tribunal of public candour and criticism, having experienced its kind indulgence; and being encouraged by its reception of my first essay to hope, that my present labours may not injure the reputation its former fostering kindness has raised, and its avowed sanction has confirmed.—Glowing with the warmest effusions of gratitude to my numerous friends and subscribers, and more particularly to those eminently learned men who have so liberally championed my appeal to the public by their distinguishing friendship and suffrage, I offer them the emanations of my grateful sensibility; which, like the diffused rays of the sun, convey but feeble impressions of that warmth—which, concentrated in my bosom, burns with an ardour that can never be extinguished by the chilling ungenial damps of ingratitude and forgetfulness.

M. BRYAN.

TABLE OF CONTENTS.

c

CONTENTS.

CONTENTS.

CONTENTS.

SUBSCRIBERS.

HIS ROYAL HIGHNESS THE DUKE OF SUSSEX.

A.

Her Grace the Duchess of St. Alban's.
His Grace the Duke of Athol.
The Right Honourable Lady Arden.
The Right Honourable and Reverend Lord Charles Aynsley.
Lady Astley, *Burgh Hall, Norfolk.*
J. Abbot, Esq. *Blackheath.*
F. Accum, Esq. *Old Compton Street, Soho.*
Miss Affleck, *West Farm.*
Mrs. Ansley, *Broad Street.*
H. Andrews, Esq. *Milk Street, Cheapside.*
Messrs. J. and A. Arch, Booksellers, *Cornhill,* 6 copies.
C. Arnold, Esq. *Bedford Row.*
T. Ashmore, Esq. *Thavies Inn.*
Miss Ashmore, *Gower Street.*
Miss Atkins, *Portland Place.*
Miss H. Atkins, *Bryan House.*
Sir James Alexander, *Burrows Buildings.*
Mrs. Auldjo, *Finsbury Square.*

B.

The Right Reverend Dr. Buckner, Lord Bishop of Chichester.
The Right Reverend Dr. Burgess, Lord Bishop of St. David's, *Abergwilly.*

d

The Right Honourable Countess of Bristol, *St. James's Square.*

Sir Richard Brooke de Capel Brooke, Bart. *Great Oakley, Northamptonshire.*

Honourable John Barrow.

Charles Burney, LL.D. F.R.S. Professor of Ancient Literature.

G. Birkbeck, M.D. Professor of Natural Philosophy and Chemistry, *London.*

W Bailey, Esq. 13th Regiment.

Miss Bailey, *Bryan House.*

John Baker, Esq. *North Down, Kent.*

Lieutenant-Colonel Bannerman, *Harley Place.*

Mrs. Barratt, *Stockwell.*

P. Barnes, Esq. *Surrey Place.*

C. Le Bas, Esq. M. C. *Bath.*

E. D. Batson, Esq. *Gower Street.*

Mrs Bateson, *Argyle Street.*

—— Baxter, Esq. *Dundee, North Britain.*

Miss Baxter, *Ditto.*

Daniel Beale, Esq. *Fitzroy Square.*

George Beale, Esq. *Waterford.*

C. Beaumer, Esq. *York Street.*

Miss Beaty, *Newport Pagnel, Bucks.*

George Bedford, Esq. *Bedford Row.*

Miss Bedford, *Ryegate.*

Edward Bell, Esq. *Halifax.*

Mrs. Bennet, *Cadbury House, Somersetshire.*

Arthur Bernie, Esq. *Waterford.*

R. Best, Esq. *Greenwich.*

Mr. Bettison, Library, *Margate.*

Mrs. Bevan, *York Place, Mary-le-Bonne.*

Miss Black, *Gower Street.*

Messrs. Black and Parry, Booksellers, *Leadenhall Street*, 6 copies.

Mrs. Bladen, *Twickenham.*

—— Blair, Esq. *Perth, North Britain.*

Mrs. Blair, *Dublin,* 2 copies.

Mrs. Boyd, *Margate, Isle of Thanet.*

William Boyd, Esq. *Blackheath.*

Mrs. Boyne, *Wanstead.*

John Paramour Boys, Esq. *Danbury, Essex.*

Mrs. J. P. Boys, *Ditto.*

H. Boys, Esq.

John Boys, Esq. *Salmeston,* near *Margate.*

Major Brace.

Miss Bradley, *Gore Court, Woodstock.*

Miss E. Braithwaite, *Greenwich.*

Mrs. Breese, *Crutched Friars.*

Mrs. Brograve, *Springfield Place, Essex.*

Miss Broadreth, *Liverpool.*

William Brown, Esq. *Green Street, Grosvenor Square.*

Mrs. Bryan, *Eccles Street, Dublin,* 2 copies.

Miss E. Buckworth, *Tottenham.*

Mrs. Burgess, *Brook Farm, Hants.*

W. R. Burgess, Esq. *Strand.*

Miss Burrell, *Southall, Middlesex.*

C.

Mrs. Cairncross, *Pancras Lane.*

David Carruthers, Esq. *Norfolk Street.*

Mrs. Cawne, *Mercers' Hall.*

Miss E. Cawne, *Ditto.*

Major Christie, *Blackheath.*

John Christie, Esq. *Ditto.*

Miss Christie, *Ditto.*

John Chuter, Esq. *Upper Homerton.*

R. Clark, Esq. Chamberlain of the City.

Mrs. Clark.

R. Clark, Esq. Jun.

John Clark, Esq. *Christ's College, Oxford.*

Miss Clark.

—— Cope, Esq. *Fenchurch Street.*

J. C. Coleman, Esq.

Mrs. Compton, *Booth Street.*

Mrs. Coneybeare, *Blackheath.*

Colonel Cookson, *Royal Artillery.*

Miss Cook, *Green Street, Grosvenor Square.*

Joseph Cotton, Esq. *Laytonstone.*

Mrs. Cowell, *Margate.*

Mr. Crosby, Bookseller, *Stationers' Court, Ludgate Hill,* 6 copies.

Reverend W. Cruttenden, *Bury.*

D.

The Right Honourable Lady Dacre, *Lee, Kent.*

Sir John Douglas, *Greenwich Park.*

The Dean and Chapter of Durham's Library.

F. Damiani, Esq.

Mrs. Davison, *Great Ealing, Middlesex.*

Crawford Davison, Esq. *Throgmorton Street.*

Mrs. Davison, *Hornsey.*

John Davies, Esq. *Cork Street, London.*

S. Davies, Esq. *Borough.*

Mrs. T. Dawson, *Jeffries Square.*

Mrs. Dawson, *Kentish Town.*

J. Deane, Esq. *Winchester.*

Mrs. De Champs, *Blackheath.*

Kennet Dixon, Esq. *Throgmorton Street.*

Samuel Douglas, Esq. *America Square.*

Mrs. Douglas, *Orchardton, Galloway, North Britain.*
—— Dunbar, Esq. *Upper King's Street, Bloomsbury.*
Reverend John Dubourdieu, *Anahilt, Ireland.*
Lieutenant Dubourdieu, *Royal Engineers.*
Lieutenant-Colonel Ducket, 2 copies.
Mrs. Duncan, *Blackheath.*
Mrs. Dunlop, *South Street, Finsbury Square.*
George Dwyer, Esq. *Blackheath.*

E.

The Right Honourable Countess of Euston.
T. Edmeads, Esq. *Greenwich.*
John Edwards, Esq. *Blackheath.*
P. Eudo, Esq. A. M. *Cursitor Street.*
Mrs. Evans, *Horton Cottage, Berks.*

F.

Honourable Colonel Frazer, *Lovat, North Britain.*
Mrs. William Falton, *Watling Street.*
Mrs. Farrell, *Dublin.*
Mrs. Favenc, *Russel Square.*
Mrs. Fysche, *Camberwell.*
Miss Flower, *Finsbury Square.*
Mrs. Foudrinier, *Guildford Street.*
Miss E. Fox, *Wood Cottage, Falmouth.*
John Furtado, Esq. *Walham Green.*

G.

Her Grace the Duchess of Grafton.
Francis Garratt, Esq. *London Bridge.*

SUBSCRIBERS.

W A. Garratt, Esq. *Trinity College, Cambridge.*
Miss Goodwyn, *Blackheath.*
Mrs. Gooch, *Brunswick Square.*
F. Gosling, Esq. *Bloomsbury Square.*
Mrs. Graham, *Liverpool.*
Miss Grant, *Upper Baker Street.*
Miss E. Grant, *Bryan House.*
Joseph Green, Esq. *Guildford Street.*
O. G. Gregory, Esq. *Royal Military Academy, Woolwich.*
Mrs. Groombridge, *Blackheath.*
Mrs. Grove, *York Place, Portman Square.*

H.

Right Reverend Dr. Horsley, Lord Bishop of St. Asaph.
—— Hall, Esq. Royal Navy.
Sir Busick Harwood, M.D. F.R.S. F.A. Professor of Anatomy, *Cambridge.*
Miss Hall, *Renishaw, Derbyshire.*
William Stewart Hamilton, Esq. *Dublin.*
—— Harding, Esq. *Pall Mall.*
C. Harvey, Esq. *Essex Street.*
Mrs. Harvey, *Finsbury Square.*
Miss Harvey, *Bryan House.*
Mrs. Harvey, *Sandwich.*
Mrs. Havelock, *Ingress Park*, 3 copies.
Miss Hawkins, *Twickenham.*
Miss E. Haynes, *Twickenham.*
Colonel Thomas Hickshaw, *Cumberland Place.*
John Hicks, Esq. *Bartholomew Close.*
—— Hindes, Esq. Treasurer to the Island of Trinidad.
James Holcroft, Esq. *Stepney.*

Joseph Holden, Esq. *Lombard Street*.

Miss Houghton, *Reading*.

Thomas Hutchson, Esq. *Walbrook*.

—— Howell, Esq. *Pall Mall*.

C. Hutton, LL.D. F.R.S. Professor of Mathematics in the Royal Military Academy, *Woolwich*.

Colonel Hutton, Royal Artillery.

Mrs. Hutton.

Miss Hutton.

Miss E. Hutton.

I.

C. Idle, Esq. *Adelphi Terrace*.

—— Idle, Esq. *Strand*.

James Innes, Esq. *Layton, Essex*.

Miss Jameson, *Roehampton*.

Mr. Johnson, Bookseller, *St. Paul's Church Yard*, 6 copies.

Mrs. E. K. Jones, *Walthamstow*.

K.

Mr. Kearsley, *Fleet Street*, 50 copies.

Mrs. G. Kearsley.

Mrs. Kearsley, *Margaret Street*, 2 copies.

P. Kelly, Esq. *Finsbury Square*.

Miss Kenrick, *Chilham*.

Thomas Kesteven, Esq. *West End, Hampstead*.

Joseph Kesteven, Esq. *Ditto*.

Mrs. Kindersley, *Blackheath*.

Mrs. N. Kindersley, *Ditto*.

Reverend G. King, *Trinity College, Cambridge*.

S. King, Esq. *Waterford*.

Mrs. Knox, Lady of the Bishop of Derry.

L.

The Right Honourable and Reverend Dr. Legge, Dean of Windsor.

Lackington, Allen and Co. *Finsbury Square,* 6 copies.

William Laithley, Esq. *Hackney.*

Mrs. Larkin, *Clare House, Malling,* Kent.

Miss Larkins, *Blackheath.*

William Ashby Latham, Esq. *Hampstead.*

T. Latham, Esq. *Bexley.*

H. Ledger, Esq. *Blackheath.*

Messrs. Leigh and S. Sotheby, Booksellers, *Strand,* 6 copies.

—— Litt, Esq. *Liverpool.*

Mrs. Leverton, *Bedford Square.*

Messrs. Longman and Co. Booksellers, *Paternoster Row,* 8 copies.

C. Lyford, Esq. *Basingstoke.*

—— Lytton, Esq. *Ramsgate.*

M.

The Honourable Lady Milbanke.

The Honourable Miss Manners, *Pall Mall*

Reverend Dr. Maskelyne, Astronomer Royal.

Mrs. General Maclean.

R. McKennel, Esq. *Watling Street.*

Henry Stowe Man, Esq. *Kennington.*

Miss Man, *Bryan House.*

H. Man, Esq. *Harp Lane.*

Miss Mansell, *Lodge, Trinity College, Cambridge.*

Mrs. Provost Marshall, *Perth, North Britain.*

Mrs. Marryat, *New Bridge Street.*

R. Mangles, Esq. *Ealing.*

James Maxwell, Esq. *Trinidad,* 2 copies.

J. P. Meyer, Esq. Jun. *Newman Street.*

—— Miller, Esq. *Dundee.*

S. G. Mills, Esq. *Greenwich,* 2 copies.

Mrs. Masson.

H. I. De Michele, Esq. *Charlotte Street, Portland Place.*

Mrs. Mitchell, *Portland Place.*

—— Mitchell, Esq. *Dundee.*

David Masson, Esq. *Perth.*

J. G. Morgan, Esq. *St. John's College, Cambridge.*

John Morrice, Esq. *Malling, Kent.*

—— Morrison, Esq. *Perth.*

Mrs. Munro, *Blackheath.*

Daniel Mussenden, Esq. *Larchfield, Ireland.*

Miss Monins, *Canterbury.*

N.

His Grace the Duke of Northumberland.

J. Nash, Esq. *Wakingham, Berks.*

The Right Honourable Sir John Newport, Bart. Chancellor of the Exchequer of Ireland.

William Newport, Esq. *Belmont, Waterford.*

Reverend F Newport, *Ditto.*

Captain Newport, 8th Regiment.

Samuel Newport, Esq. 58th Regiment.

M. Newport, Esq. *Belfast.*

Miss Newport, *Bryan House.*

Mrs. General Nicholson, *Upper Baker Street.*

R. Nixon, Esq. *Doughty Street.*

Josias Nottidge, Esq. *Bocking, Essex.*

Thomas Nottidge, Esq. *Ditto.*

William Nottidge, Esq.

J. T. Nottidge, Esq.

O.

Sir George Mulgrave Ogilvie, Bart. *Barras, North Britain.*
Sir H. Oxenden, **Bart**. *Broome, Kent.*
Lady Oxenden.
R. O'Conner, Esq. *Dublin.*
William Ogg, Esq. *Chiswell Street.*
Mrs. Oldfield, Ladies Seminary, *Margate,* 2 copies.
Mr. Ostell, Bookseller, *Ave Maria Lane,* 6 copies.

P.

The Right Reverend Dr. Percy, Lord Bishop of Dromore.
The Honourable Spencer Perceval, Attorney-General.
Lady Prescott, *York Place, Portman Square.*
J. G. Palairet, Esq. *Bath.*
Edmund Paley, Esq. *Queen's College, Oxon.*
Signor Parachinetti, 2 copies.
J. Parkinson, Esq. *Hampstead.*
Reverend G. Paroissien, *Hackney.*
Miss Papillion, *Lee, Kent.*
Mrs. Patterson, *York Place, Portman Square.*
Miss Pease, *Hull.*
Reverend H. Penny, *Ealing.*
Mrs. Perdew, *Deal.*
Leonard Phillips, Esq. *Vauxhall.*
E. Phillips, Esq. *Scotland Yard.*
J. Pickford, Esq. *Baker Street.*
Miss Pickford, *Market Street, Herts.*
Miss Pickford, *Poynton House, Macclesfield.*
Mrs. Picklington, *Sloane Street.*
E. Powell, Esq. *Preesquance, Shropshire.*
Miss Prestwidge, *Mincing Lane.*

Mrs. G. Prestwidge, *Dulwich.*
G. Pryme, Esq. *Lincoln's Inn.*

R.

His Grace the Duke of Rutland.
Miss Radley, *Bryan House.*
John Radley, Esq. *Fleet Street.*
—— Rapier, Esq. *Francis Street.*
William Rawson, Esq. *Hill House, Halifax, Yorkshire.*
Miss Read, *Bryan House.*
—— Reina, Esq. *Royal Navy.*
T. Redish, Esq. *Bank of England.*
—— Richards, Esq. *Bow.*
E. Rice, Esq. *Dover.*
Messrs. Richardsons, Booksellers, *Cornhill*, 6 copies.
Mrs. Rideout, *Court Lodge, Sussex.*
Reverend Mr. Richards, *Farnborough, Hants.*
E. Robertson, Esq. *Beverley, Yorkshire.*
Miss Robertson, *Bryan House.*
Christopher Rolleston, Esq. *Watnal, Nottinghamshire.*
Mrs. Rolleston.
Mrs. Rowly, *St. Neots, Hants.*
J. C. Ruding, Esq. *Francis Street, Bedford Square.*

S.

Right Honourable Lady Seaforth.
Lady Stuart, *Brighton.*
Lady Scott, *Great Barr Hall, Staffordshire.*
Josias Saville, Esq. *Bocking, Essex.*
G. Samuel, Esq. *Richmond Buildings.*
—— Sanderman, Esq. *Perth, North Britain.*

Miss Sawkins, *Margate.*

Miss Scatcherd, *Hull.*

Captain Schuyler, *Flushing, Cornwall.*

Miss Schuyler, *Flushing, Cornwall.*

Miss Scott, *Margaret Street.*

J. Scratten, Esq. *Hackney.*

—— Sealy, Esq. *Lisbon.*

Mr. Seeley, Bookseller, *Ave Maria Lane,* 6 copies.

William Sharp, Esq. *Ramsgate.*

Miss Sharp, *John Street, Fitzroy Square.*

Miss Short, *Upper Brook Street.*

Benjamin Shaw, Esq. *London Bridge.*

Mrs. Shute, *Sydenham.*

—— Simmons, Esq. *Falmouth.*

Mrs. Simpson, *John Street, Bedford Square.*

C. Simpson, Esq. *Ditto.*

Miss Simpson, *Ditto.*

Mrs. Sitwell, *Renishaw Park, Derbyshire.*

Sittingbourne and Milton Book Society.

Mrs. J. S. Smith, *Bennet's Hill, Doctor's Commons.*

Mrs. Snowden, *Sandwich.*

Mr. Sotheby, *York Street,* 2 copies.

Miss Sotheby.

N. St. Croix, Esq. *Homerton.*

Mrs. Stokes, *Great James' Street, Golden Square.*

Miss Stonestreet, *Sherborne, Dorsetshire.*

Mrs. Stringer, *Peckham, Surrey.*

Charles Studd, Esq. *Ipswich.*

Captain Sutton, *Windsor Castle.*

Mr. H. D. Symonds, Bookseller, *Paternoster Row,* 6 copies.

J. Strangeman, Esq. *Waterford.*

T.

Reverend Mr. Tavel, M.A. Fellow and Tutor, *Trinity College, Cambridge.*
Mrs. Tattersal, *Otterden Place, Kent.*
Mrs. Temple, *Roehampton.*
Mrs. Thompson, *Chiswick.*
W. Ticken, Esq. *Royal Military College, Great Marlow.*
W. Tiffin, Esq. *Blackfriars.*
F. Townsend, Esq. *Herald's Office.*
A. Tulloh, Esq. *Gould Square.*
Miss Tulloh.
Mrs. Twopenny, *Woodstock, Kent.*
Mrs. Osburn Tylden, *Torry Hill, Kent.*
N. C. Tyndal, Esq. *Temple.*

V.

Dr. Vaughan, *Curzon Street.*
Benjamin Vaughan, Esq. *Great James' Street.*
Messrs. Vernor, Hood and Sharpe, 6 copies.
Mrs. Vivian, *Lincoln's-Inn Fields.*

W.

Reverend Dr. White, Professor of Hebrew and Arabic, Canon of Christ-Church, *Oxford.*
Lady Wilson, *Charlton.*
Sir T. Wilson, Bart.
Reverend E. Whitby, *Lighford, Staffordshire.*
Mrs. J. Walker, *Blackheath.*
Miss Walker, *Southampton.*
Mrs. Wallace, *Charlotte Street, Portland Place.*
Reverend C. T. Waller, *Blackheath.*

Reverend John Watson, *Morden College, Blackheath.*
Mrs. Way, *Acton.*
C. Wilkinson, Esq. *Clapham.*
John Williams, Esq. Commissioner of the Customs.
Miss Williams, *Horton Cottage, Berks.*
Robert Williams, Esq. *Bedford Square.*
Mrs. Williamson, *Dundee.*
Mrs. Wilson, *Gower Street.*
R. Wilson, Esq. *Waterford.*
C. W. Wiple, Esq. *Rutland Place.*
Miss Wiple.
Mrs. Wools, *Winchester.*
T. Worthington, Esq. *Halliford.*
C. Worthington, Esq. *Ditto.*
Miss Worthington, *Ditto.*

Y.

Reverend R. Young, D.D. *Hackney.*

Z.

Miss Zurhurst.

MRS. BRYAN EDUCATES YOUNG LADIES

AT

BRYAN HOUSE, BLACKHEATH.

ERRATA.

Page 89, line 2, dele *with*.
    92, 23, dele *and*.
   176, 28, for *tube*, read *table*.
   212, 26, dele *of*, after applying.
   221, 22, dele *from the point* w.
   269, 12, for *phosphorics*, read *phosphori*.
   298, 24, for *equal*, read *parallel*.
   306, 26, for *degrees*, read *degree*.
   308, 23, for *longitude*, read *latitude*.
   308, 25, for *latitude*, read *longitude*.
   325, 6, for *latitudes*, read *latitude*.
   334, 17, introduce *and* after problem.
   342, 8, introduce *about* after than.
   343, 3, instead of *of*, read *from*.
   343, 13, for $62\frac{1}{2}$ *degrees*, read 62 *degrees*.
   343, 19, for 47′, read 45′.

PLATE I.

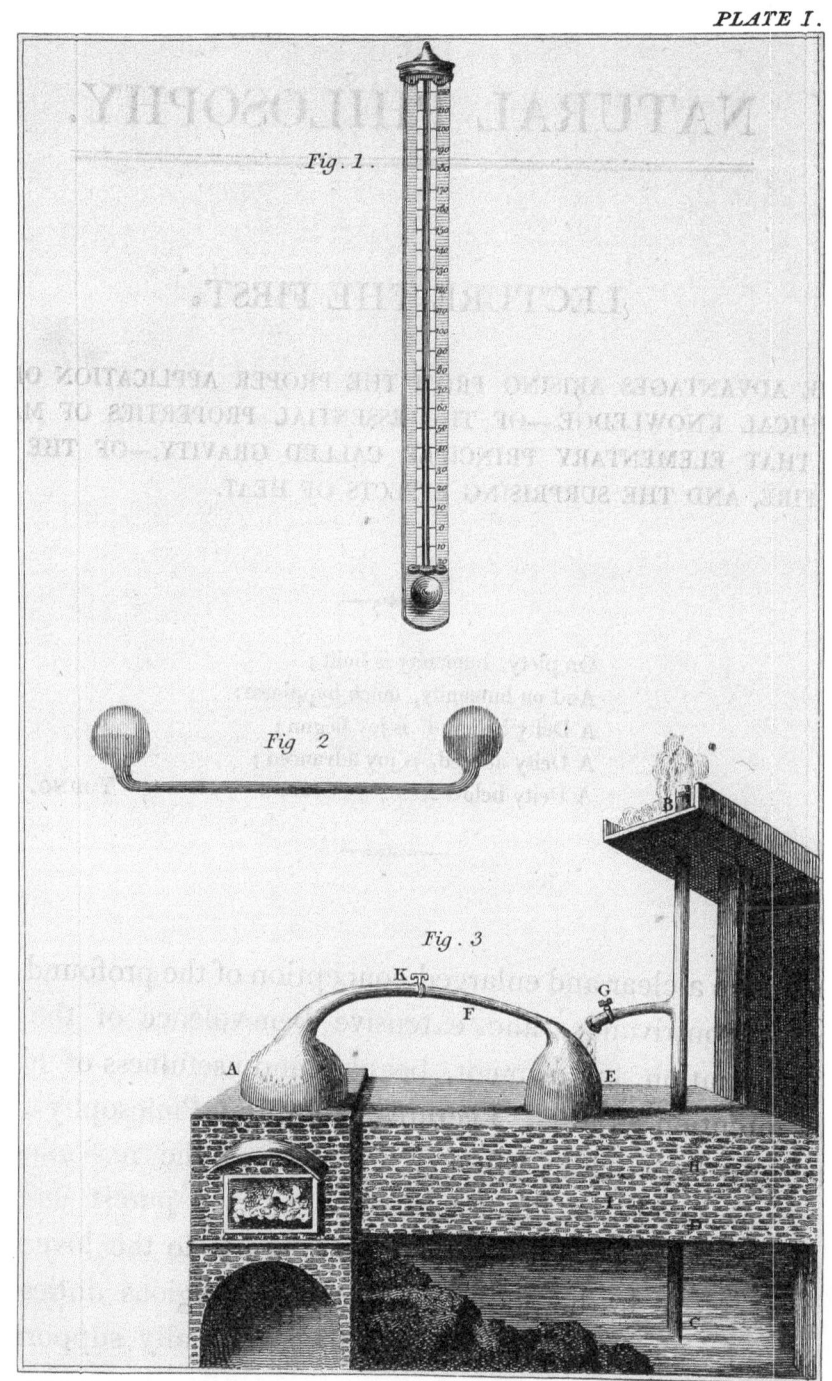

Fig. 1.

Fig. 2

Fig. 3

T. Noble delin.t

H. Mutlow sculp.t

NATURAL PHILOSOPHY.

LECTURE THE FIRST.

OF THE ADVANTAGES ARISING FROM THE PROPER APPLICATION OF PHILO-
SOPHICAL KNOWLEDGE.—OF THE ESSENTIAL PROPERTIES OF MATTER.—
OF THAT ELEMENTARY PRINCIPLE CALLED GRAVITY.—OF THE NATURE
OF FIRE, AND THE SURPRISING EFFECTS OF HEAT.

On piety, humanity is built;
And on humanity, much happiness:
A Deity believed, is joy begun;
A Deity adored, is joy advanced;
A Deity beloved, is joy matured.   YOUNG.

To induce a clear and enlarged conception of the profound wisdom, exquisite contrivance, and extensive benevolence of the Creator, in the formation, endowment, beauty, and usefulness of his works, is the intention of these Lectures. Natural Philosophy affords a most truly noble and salutary exercise of the reasoning faculties; from which the highest advantages, the purest enjoyments, instinctively arise; for it assimilates our hearts to the love of truth, and recreates us in the performance of our religious duties—Philosophy and religion being natural allies, mutually supporting and illustrating each other.

B

The objects of our contemplation in the natural world, though in one sense familiar, are yet not, perhaps, altogether so in the sense which can render them most extensively subservient to our truest interest; which the knowledge of their efficacy, truth, and sublimity can alone effect: for though the beautiful exterior of nature raises pleasurable sensations, and astonishes us by its variety and grandeur; yet a superficial observation of the wonders of creation cannot excite those delightful emotions which arise from the benevolence of the motive when understood; or that evidence of the wisdom employed in producing the effects perceived by us, which naturally results from the demonstration of the excellency, simplicity, and energy of the works of God; procurable only by that touchstone of physical truth—experimental demonstration.

OF MATTER.

In treating of Matter, we will consider its general essential properties, prior to contemplating its separate elementary ones. Matter is the substance which is the foundation of all bodies; and that, by means of our senses, renders evident all other properties which combine with it.

The essential properties of Matter are solidity, extension, and inertia. In Physics, solidity expresses the substance which contains the elementary parts of all bodies, and which may properly be called Matter. Extension is that property of Matter determined by its occupying space; and inertia that which causes it to remain at rest when not acted upon by some extraneous power to put it in motion. This power continues the motion, when a body receives a given impulse, and resists an impressed force: hence, resistance is also a characteristic of matter.

Gravity is a property of Matter, of which we know very little but by its effects. Of the invisible agent which produces the effects we call Gravity, the great and expansive genius of Newton has furnished us with the most magnificent and certain evidences that the most exalted human reason could produce : yet he was not able to discover the cause which produced the effects he perceived of it, no more than his predecessor Galileo; who first observed the tendency of this powerful agent in nature. In my lectures on Astronomy *, its chief energies are displayed; and I have shown, that the whole system of the universe, with the motions producing universal harmony, results, under the direction and impulse of Divine Providence, from the operations of this powerful principle; combined with the laws of projectile motion, and the influences of magnetism, fire, and electricity.

Gravity, cohesion, magnetism, and electricity, are incidental properties of matter; which may all be expressed by the term Attraction. In this Lecture, we will confine ourselves to the consideration of the two former properties, cohesion and gravitation. Cohesion is the attraction between bodies that keeps them together, when in immediate contact with each other; and also unites the parts of bodies, and forms them into masses. If two pieces of cork, of equal size, be placed on a surface of water, near each other, they will approach by gravity or attraction; and, agreeably to the laws of motion, with an accelerated velocity, until they meet; when, if we move either piece, the other will follow it, by the attraction of cohesion. If the pieces of cork be of unequal

---

* For a fuller illustration of this subject, I beg to refer the readers to my Lectures on Astronomy, octavo edition, in which they will perceive that impulse is the primary cause of all motion, a divine agent being implied; without which, no motion could be either generated or continued.

weight, the heavier will move slower than the lighter one; that is, their motions will be in the reciprocal proportion to their weights; the motion of each being in the inverse proportion of the quantity of matter in each body.

If two pieces of cork be placed nearer to the edge of the bason than to each other, they will be attracted by the bason, and drawn from each other, close to the edge of it. By the power or force of gravity, all bodies, when free to move, fall towards the centre of our earth; which force causes them to press the hand that stops their progress, and makes them fall in lines perpendicular to the horizon when suffered to proceed. Some of the ancient philosophers, being unacquainted with the laws of gravity, and perceiving that some light bodies rose in the air, thought there were two principles in bodies: one, by which lighter bodies tended upwards from the earth; and another, by which heavier ones descended towards it.

It was Galileo who discovered that all bodies fell towards our Earth; and that it was the different specific gravities * of the bodies, with the medium through which they passed, that caused the difference in their descent; for in an unresisting medium, he perceived they fell towards the Earth with equal velocity. This assertion we shall prove by experiment, when we treat of the resistance and other properties of Air.

It has been long disputed, whether there be a property of repulsion as well as of attraction in bodies; which, perhaps, we are

* Specific gravity is the difference of weight between two bodies of equal size, but of different densities.

not able to determine: for I conceive the effects would be the same, either by an attractive power ceasing, or a repulsive power beginning to act. But if I were to obey the dictates of my reason, I should assert, that repulsion was not an universal principle, or its effects would be more evident, or at least as evident as those of attraction, by repelling bodies with a given velocity; which is not perceived, excepting in particular instances; and then it generally results from new and distinct agents, namely, from the properties of magnetism, fire, and electricity.

The attraction of cohesion is circumscribed within narrow bounds; a very small intervening body preventing its effects: for although two drops of quicksilver will run together and form one large drop; yet, if these drops be rolled in the finest dust, the attraction will cease. Though I do not consider repulsion to be an universal principle of matter, yet I am convinced that there is a natural repulsion between some bodies; as, for example, it must be repulsion that causes needles to rest on a surface of water; for the materials are heavier than water; therefore, according to the laws of specific gravity, they would sink in that fluid: we also observe round the needles a space free from water; this can no way be accounted for but by repulsion.

OF FIRE.

The agency of Fire is so universal, and its importance in the natural world so very great, that it was the first object which excited curiosity in the inquisitive minds of natural philosophers. Various have been their opinions concerning it, the most rational of which I have adopted in these Lectures; and will endeavour to convince you of their validity in the course of our investigations.

Fire is of a subtile, active, and elastic nature, universally disseminated through the universe, penetrating all bodies, and having
a tendency to diffuse itself uniformly in all substances; dilating
every substance it penetrates; making some of them assume the
state of fluidity, and afterwards that of vapour.

It is by our senses we derive our ideas of heat and cold; yet
are they insufficient to determine the precise degrees: for the
different state of heat or cold on the surface of our bodies will
produce a weaker or a stronger sensation of either, than belongs
to the real state of fire in the body touched. Philosophers, perceiving this fallacy of the senses, naturally sought for some intermediate agent, to enable them to estimate the actual state of
heat and cold in bodies. By their knowledge of the property of
heat to expand all bodies, both solid and fluid, and of cold to
condense them, they constructed the thermometer; and nothing
could have been contrived more simple and correct for effecting
their purpose.

This instrument, fig. 1, plate I, ascertains the degrees of heat
and cold, by its power of exhibiting these varieties, and serves to
show, that fire penetrates the hardest substances; for, by applying
the tube to different bodies, we perceive that fire penetrates them
all, and has a constant tendency to place itself in equilibrium.

Quicksilver, being a fluid of a cleanly and unchangeable nature,
is used in this instrument; and, by its expansion and contraction,
shows the state of heat in any body contiguous to it. The ball
and part of the tube are filled with this fluid; the expansion and
contraction of the quicksilver are evident, by the rise and fall of it
in the tube. In order to obtain two fixed points, that the inter-

mediate divisions might be marked, it was necessary to find two extremes that would be uninfluenced by either climate or season; under which conviction, philosophers made various experiments; and at last discovered, that the temperature of boiling water, and that of freezing water in its most immediate approach to the state of solidity, were always in the same degree under each circumstance. Thus furnished with the means of effecting their purpose, they plunged the ball of quicksilver into melting ice, and observed the fall of it in the tube; and afterwards into boiling water, and remarked its expansion or rise in the tube: from these two fixed points they could regulate the intermediate degrees with certainty; which they marked exactly in the proportion that the different states of heat and cold raised and depressed the quicksilver in the tube. Fahrenheit's thermometer is the one chiefly used; and his scale is thus expressed: The freezing point is marked 32°, being that at which the quicksilver rests, on being immerged in freezing water; and the boiling point 212°. Between these two points there are 180 intermediate degrees. Why 0 was placed beneath the freezing point is not known; but we may suppose it was to allow for the greatest degree of cold that is produced on our Earth. We know the cold must be much greater at the poles than with us; and even in these latitudes we often experience cold considerably greater than that of water in a freezing state.

That Fire elevates the quicksilver in the tube we are certain, by applying a taper to it; but whether it acts by penetration or by contact is not easy to determine. It may impart some property which causes expansion, without actually infusing any particles to extend its bulk; for a thermometer weighs no more when heated than when cold: therefore, as Fire is a material substance, we cannot readily subscribe to the idea of an increased quantity in the tube; but may more easily conceive, that the natural fire in the tube is rendered active by the impression it receives, and raises the quicksilver by its increased elasticity; yet, as we know that Fire penetrates all substances, I think it also probable that it penetrates and passes through the tube; and, by its rapid motion, communicates the property of expansion, or elasticity.

That light and fire are emanations of the same agent is evident, by observing the effects of the solar light on the surface of a convex lens, which, when condensed at its focus, exhibits pure fire. That light is a material substance we know, for we can decompose or divide its parts; therefore must fire be also material. Numerous are the characteristics of substance which Fire exhibits; but what I have said must be sufficient to convince you of its materiality; as also of the identity of light and fire.

We find, by experience, that Fire resides in substances cold to the sense, and is not perceived till rendered active by the dilatation of the particles of bodies; as, for example, steel, which is cold to the touch, contains fire in a quiescent and latent state, as is rendered evident by striking it against a flint, a substance harder than itself; from which we infer that heat, the effect of Fire, is not perceived but when in an active state. The action of Fire may be either free or confined. When its activity is augmented to a great degree, it will burst from the form which confines it, however compact or hard its nature; and the energies of this agent are so powerful, that it will destroy all substances, when rendered considerably more active than in its natural state in each of them.

As we proceed, we shall perceive how this universal constitutional medium of all natural bodies, notwithstanding its power to consume all things, is kept in subordination by the wisdom of God, and rendered the chief agent in all the operations of nature; and that without it the animal, vegetable, and mineral kingdoms— those beautiful emanations of divine wisdom and goodness—would be no more.

Fire has a constant tendency to place itself in equilibrium; which may be accounted for by its natural activity, which endeavours to displace all impediments to its progress. Let us in like manner exert our energies, to overcome all impediments to our progressive mental attainments. That knowledge which is the most essential to our happiness, and the perfection of our nature, requires the most active exertions to obtain: let us apply with due force to effect our purpose; and when possessed of the essential parts of religious, moral, and physical truths, endeavour to expand each subject, and communicate our attainments for the benefit of our fellow creatures. This elevation of mind, and glow of benevolence, like the effect of heat on our atmosphere, will raise our genius, and extend our influence; for to enforce the practice of piety by the contemplation of natural objects, is to keep the mind constantly renovated in the contemplation of its excellence, importance, and indispensability: by which our taste becomes refined; our influence diffused, forcible, and permanent; and our charities extended towards the whole race of created beings. That it is the nature of heat to dilate the air of our atmosphere, is made evident by a very simple experiment. Place the flame of a candle at the top of a door-way, in a room heated by the sun or a fire, and the flame will bend outwards from the room towards the passage or hall: on removing the candle to the bottom of the door-way, the flame will bend inwards to the room; for the heated air of the apartment is expelled out at the upper part of the room, by the colder air rushing in at the lower

c

part of it.   If the candle be held in the middle of the door-way, the
flame will not be at all affected; because the chief current of air
is at the top and bottom of the apartment, which current decreases
regularly towards the centre.   The heated air rises to the top, as
it is less resisted in that than in any other direction.

The property of transmitting heat differs, according to the substances the fire penetrates or resides in.   Metals convey fire very
fast: for on presenting one end of a metal rod to the fire, and holding the other in the hand, it soon becomes too hot to be held; but a
rod of glass may be held a much longer time without inconvenience.
All porous bodies convey fire slowly—such as wood, cork, and spun-
gy substances; probably because there is less contact of the parts
that compose them: for we perceive more powerful effects of the
transmission of fire, in substances of a close than in those of a loose
texture; so that the latter must retain heat better than the former.
Woollen cloth keeps our bodies warm, by preventing the escape of
our constitutional heat: snow is particularly serviceable to our earth,
for, being a covering of a loose texture, it prevents the evaporation
of the internal heat of the earth, and thereby assists the business of
vegetation: hence the great wisdom and benevolence of our Creator
and Preserver, in clothing the surface of the earth in cold climates
with an insulating substance, which prevents the dissipation of the
constitutional heat necessary for the growth of vegetables, and the
support of animal nature.

Water conveys fire very fast; yet deep lakes of water are seldom
frozen; for unless the frost be of very long duration, the whole heat
of such large portions of water cannot be carried off by it.

Whatever be the state of heat peculiar to individual animals, they preserve it independently of external circumstances, while in a healthy state; being provided with the means of throwing off superabundant heat by the pores; and of producing and retaining their internal heat, by the various operations carrying on within them.

OF EVAPORATION.

HAVING mentioned the ability of the animal body to throw off superabundant heat, we will proceed to consider the subject of evaporation in general. Heat causes bodies, both solid and fluid, to rise in vapour; and that this evaporation discharges heat from them, is certain, by a variety of familiar evidences. Of the effect of heat on water in producing vapour, we have an example when a small quantity is placed in a vessel on the fire; for as soon as it becomes hot it evaporates, and carries off a portion of the fluid with it: we perceive the same effect of fire on a candle, the flame carrying off the whole substance of it in vapour. The substances of some solids evaporate faster than others: the lighter go off with a weaker heat than those which are heavier; those bodies which are extremely light are called volatile.

The cooling property of vapour is evident by the observation of its effects, and the manner in which they are employed in hot countries. In Aleppo, water kept in jars is always coolest when the weather is hottest: for when the heat is most excessive, and the sun's rays most powerful, the vapour from the outside of the jars is most copious; and the degree of coldness within them is produced by the great quantity of heat discharged through the pores of the earth of which they are made, which is of a very loose texture. The manner of obtaining ice in the East Indies is another evidence of the degrees of cold that may be produced by evaporation. The ice-makers dig pits about thirty feet square and two feet deep, in large plains, strewing the bottom of these pits with sugar canes: they place upon them unglazed pans, made of such porous earth that the vapour penetrates through its substance. The pans are about a quarter of an inch thick: if they are filled in the evening with water that has been boiled, and left in that situation till morning, more or less ice will be found in them, according to the temperature of the air, there being more formed in dry weather than in cloudy. We may imagine that the great power of the sun in the day-time in those regions, prevents the coldness that would otherwise be produced by evaporation. A strong current of air is also necessary; for it is only by a continual influx of dry air that the whole heat of the water can be carried off: for the air becoming saturated with the vapour will receive no more; but a fresh portion continually coming in contact with the evaporating surface absorbs it, and thus deprives the water of all its constitutional heat.

The animal body is provided by God with a covering, of which the parts are capable of a dilatation necessary to suffer the superabundant heat, and impure effluvia, to evaporate through its substance. So important is this operation to animal existence, that were not the noxious vapours and superabundant heat thus to escape, they would act on the internal organs of the body, and destroy the animal functions; as is the case in fevers, which can no way be so effectually relieved, as by a copious discharge at the pores. Many of the occupations of mankind could not be carried on, but by the power of increasing the discharge at the pores, in proportion to the increased degree of heat endured from labour, or from external circumstances; and in order to supply the diminution of strength, the persons employed in them, drink strong beer, which both stimulates the action of the pores and supports nature: by these means the heat is rendered supportable under the exertions of labour and all other greatly exciting causes.

A quick current of warm dry air is wholesome, as it carries off damp vapours; and hence it is that leaving the doors and windows of a house open in summer, cools and refreshes the apartments. The great power of vapour when confined, has been exhibited in various ways, and applied for effecting some of the most difficult processes in the mechanic arts.

As we may not have an opportunity of seeing the grand effects of the operations of vapour, which are applied only on particular occasions, when great power is required; we will contemplate its simple effects, from which we may form some idea of its combined energies. The balls, fig. 2, pl. I, are partly filled with coloured water: when I hold one of them in my hand it becomes warm; and the greater the degree of heat, the sooner will the ebullition take place, and the greater force the vapour will exert. The action of heat being confined in the balls, exerts the greater energy; and so carries the fluid along with it. I will hold one ball in my hand, and you will see the fluid it contains pass into the other ball by the force of the vapour.

Two things may be understood from this instrument;—that water will boil much sooner, and with less heat, when not subjected to the atmospheric pressure, and the cold produced by evaporation: for the water boils in a moment, and flies off from the hand by the increased force of the vapour; and while the evaporation continues, or while any fluid remains in the ball held in the hand, it becomes colder and colder; proving that the greater degrees of cold are produced by more and more of the heat being carried off from the residue that remains in the ball held in the hand.

The effects of steam when employed in the machine called a steam-engine are so powerful and of so great utility, that I think the consideration of one used to raise water may be entertaining and instructive; particularly as many of us may never have the opportunity of ocular demonstration of the amazing power of steam, when increased to a very high degree. Captain Thomas Savery erected a steam-engine for the York Buildings Company in London, for the purpose of supplying the inhabitants of the Strand, and its neighbourhood, with water.

This machine is represented fig. 3, pl. I. A, represents a copper boiler placed on a furnace: E is a strong iron vessel communicating with the boiler by means of a pipe, F, at top; and with the main pipe by a pipe, 1, at bottom. BC is the main pipe immersed in the water at C; D H are the places of two valves, both opening upwards; one being above and the other below the pipe of communication, I*: lastly, at G is a cock, which serves occasionally to wet and cool the vessel E, by water from the main pipe. K is a cock in the pipe of communication between the vessel E and the boiler A.

This engine is set to work by partly filling the copper with water, and also the upper part of the main pipe above the valve H; the fire in the furnace being lighted at the same time. When the water boils strongly, the cock K being opened, the steam rushes into the vessel E, and expels the air from thence through the valve H. The

---

* A valve is a piece which lifts up in one direction, but will not open in the opposite one; thus permitting a fluid to pass out of a place, and preventing its return to it again.

vessel E thus filled, and violently heated by the steam, is suddenly cooled by the water falling upon it from G. To prevent any steam from passing from the boiler, K is at the same time shut : hence the steam in E becoming condensed, leaves the cavity almost a vacuum ; and therefore the pressure of the atmosphere at C, forces the water through the valve D, till the vessel E is nearly filled. The condensing cock, G, is then shut, and the steam cock, K, opened ; when the steam rushes again into E, and expels the water through the valve H, as it before did the air. Then E again becomes full of hot steam, which is cooled and condensed as before. The water rushes through D by the pressure of the atmosphere at C, and E is again filled ; which water is forced into the main pipe B C. Thus it is easy to conceive, that by this machine water can be continually raised as long as the boiler continues to supply the steam. So great is the expansive quality of steam, that water may be rarefied 14000 times when it is violently heated ; but it requires amazing strength in the vessel to contain it in its rarest state.

The steam raised from common boiling water is nearly 3000 times rarer than water, and almost four times rarer than common air. The experiments on steam are inconvenient, and sometimes attended with great loss and danger : the following instance of which happened on the sudden application of great heat to a small quantity of moisture. The workmen at a cannon-foundery in Moorfields running the hot metal into a mould, which had a small quantity of water at the bottom, the vapour burst the foundery to pieces.

The steam-engine is of the most extensive utility. But for this machine we could never have enjoyed the advantages of coal for

fuel in our time; as our forefathers had dug the pits as far as they
could go: for the water rising in them from different fissures prevented working deeper; but this machine enables us to drain all
the water from the mine, as it rises to impede the miners in their
progress.

Various are the uses and application of this machine: but
the advantage above mentioned is sufficient to make it interesting to us; for we are all sensible of this benefit derived from its
operations.

The effect of fire which we call evaporation is concerned in the
constitutional existence of the earth, and all it produces on its
surface. The spontaneous perspiration, or vapour, from the surface
of the earth, is not perceived, any more than that of our bodies, but
when much increased; therefore, although it is continual, yet we
cannot say it is familiar.

Experiments have been made to ascertain the quantity of spontaneous evaporation from the earth; which though we cannot suppose that they can prove, by infallible calculation, the real quantity
that is actually raised from a given portion of the earth, yet, making
proper allowances for the difference of seasons, soils, and atmospheric pressure, they will enable us to form some estimation on the
subject; at least, they will serve to establish the truth of one thing,
namely, spontaneous evaporation itself. Experiments on this subject were made by a very learned man, Dr. Watson, bishop of Llandaff; who performed them with great nicety, and observed them
with accuracy: yet we may conceive he did not mean to establish
the idea, that the actual quantity raised from a given portion of the

earth in the natural way, could be accurately ascertained by this artificial mode.

He placed a drinking glass, with the mouth downwards, on a grass-plot mowed close, and when no rain had fallen for more than a month: in two minutes he perceived the vapour rising in the glass: in half an hour drops of water trickled down the sides. This operation he repeated several times with equal success. In order to estimate the quantity thus raised in a given time from a certain portion of earth, he measured the area of the mouth of the glass, which proved to be twenty-four inches. There are 1296 square inches in a square yard, and 4840 square yards in an acre; therefore he inferred, that if from twenty inches of ground, in one quarter of an hour, the quantity raised could be ascertained, it would show how much would be raised from an acre in that time. In order to effect this, he suffered the glass to remain a quarter of an hour, and then, on removing it, wiped the inside of it with a piece of muslin which he had previously weighed. On repeating the operation several times with the same glass, the same muslin, and through the same space of time, he found, on an average, that the vapour raised in a quarter of an hour from twenty square inches of earth, was six grains. Agreeably to this calculation, he inferred, that from an acre of ground, in twelve hours, about 1600 gallons of water might be raised; but, of course, he did not mean to say he could ascertain the exact quantity raised from it by these means: for doubtless an increased quantity is sometimes raised by the heat of the sun and other causes.

Vegetables perspire, and the vapours from the earth, and the substances on its surface, are supported in the regions of air, and fall down in that pure state; and thus the constitution of the earth, and of its inhabitants, is kept in a healthy state, by the wise dispositions of nature. The greater part of the inward heat of our bodies escapes in vapour from our lungs, in an impure state; which makes a close room in which many people are assembled unwholesome, unless properly ventilated. That cold condenses air, and that air contains moisture, is evident on bringing a glass of cold water into a room where many people are assembled; for the hot air endeavours to place itself in equilibrium on the water, by which means it becomes condensed on the outer surface of the glass; and the watery particles contained in the air show themselves on the surface of the glass, being regenerated from a state of vapour into water by condensation.

In frosty weather we perceive the same effect on the windows of a warm apartment; the hot air, endeavouring to place itself in equilibrium, is pressed against the glass, by which condensation the humid particles of vapour in the air appear on its surface. Stone pavements, and some walls within-side a house, look damp after extreme cold weather; even when there has been neither rain nor snow. The reason is obvious; the fire in the substances being diminished during the cold, on the return of warmer weather the fire resident in the air endeavouring to restore the equilibrium, the humid particles of the air become condensed on these surfaces.

That the atmosphere is produced from the earth, and the substances on its surface and within it, is evident by the animal, vegetable, and mineral substances, which are employed chemically to produce different kinds of air.

That different substances absorb fire, in a greater or less degree, and that the power of imbibing heat from the sun's rays, depends

on the colour of the substances employed, is evident by the observations made by Dr. Franklin. He placed on a surface of snow,
when the sun was above the horizon, several pieces of cloth of
different colours, but of the same texture: in a few hours, the dark
colours were buried in the snow, while the white remained on its
surface; and, according to the darkness of the colour, such was
its power of imbibing heat, made evident by the pieces of cloth
sinking agreeably to the degree of that quality each possessed.
The peasants on the mountains of Switzerland spread a black earth
on the surface of snow, when they wish to melt it, in order to cultivate their lands.

The different capacity of imbibing heat in different bodies,
depends likewise on their surfaces; such as are polished and transparent, and reflect much light, not being heated in so great a degree
as those which are rough and opaque; for these, instead of reflecting light copiously, imbibe it in large quantities.

The sun is the source of heat on our globe; yet we perceive
that his rays do not heat all bodies in an equal proportion. The
greatest portion of the sun's heat is received in a secondary way
from the earth; for the atmosphere being semi-transparent, the
greater portion of the rays pass through it, and are received by
the substances on the surface of our globe, and reflected to us
from thence, unless when we are exposed to their direct influence.

The luminous appearance of heat is called ignition: the degree
of heat requisite to make bodies flame, is supposed to be according
to the texture of each body; therefore no universal standard can be
found.

Although the effects of heat and light may be separated, yet we have reason to believe that they are kindred properties, dependant on the operations of that subtle agent, fire.

The effect of fire, denominated combustion, is produced by its operations on substances capable of inflammation ; whereby they first suffer an augmentation of heat, arising chiefly from an intestine motion which produces flame, and lastly a total change in the substance burned.

Air is essential to combustion, as without air no body can emit flame. A very easy experiment will prove this assertion. Take spirit of wine, which is of a most inflammable nature, and plunge into it a red-hot coal ; and the fire will be as much extinguished as if you were to plunge the coal into water. The reason is obvious : by immersing it under the surface of the spirit, you prevent the action of air on it; but if you present a hot coal to the surface of the spirit, so that the air can act on it, the spirit will flame.

Flame is a luminous vapour raised by fire, with such an augmentation of heat, as to throw out a bright light. The degree of heat and light so thrown off from substances, is proportioned to their quantity of vaporous matter: thus, wood and coals flame violently ; whilst lead and tin scarcely emit any flame, even on the greatest increase of heat.

Smoke may be rendered luminous, by a continual influx and current of pure air to a lamp, which flames violently ; for the greater the current of air, the brisker the fire burns ; and the purer the air, the more power it exerts. If fresh air be continually supplied to a lamp, through a narrow tube, to prevent its divergence, and to concentrate its effects, they may be augmented to such a degree as to ignite smoke.

Though human nature cannot discover the means by which the Almighty performs all his wonderous works, yet those things which we are capable of investigating by the aid of our reason, confirm the truths of revelation; and, together, form the most complete confirmation of an over-ruling Providence: wise and benevolent in his decrees—great and wonderful in the operations of his hands!

As we continue our researches, we shall find in each new subject new cause of admiration; for throughout the subjects of nature we shall trace continually a directing Power: a Being independent of matter, who has designed every thing with such wisdom, and in such exquisite proportion, that nothing is either redundant or wanting; for the gifts imparted to every part of his creation, are adapted to the necessities of each. The inferior animals He has endowed with the means of preserving themselves, and a powerful natural instinct for the preservation of their offspring. To man He has imparted a stronger sense, a higher, a purer joy—the joys of intellect, the promise of eternity! How great these boons! How abundant should be our gratitude! How should we admire, adore that Benevolence that has imparted to man an immortal soul! In this lies our dignity, our superiority, of which nothing but sin and ingratitude can deprive us. Let us never suffer such baneful influences to destroy the sublimity of our nature, nor blast the blossoming expectations which in maturity will reward us with the fruition of never-ending, never-satiating delight.

———————————Thrice happy man!
And sons of men, whom God hath thus

Created in his image, there to dwell,
And worship him, and in reward to rule
Over his works on earth, in sea, in air,
And multiply a race of worshippers,
Holy and just; thrice happy if they know
Their happiness, and persevere upright.

MILTON.

PLATE II

Fig. 1.

Fig. 2.

Fig. 3.

M. Bryan design.   T. Noble delin.   H. Mutlow sculp.

PLATE III.

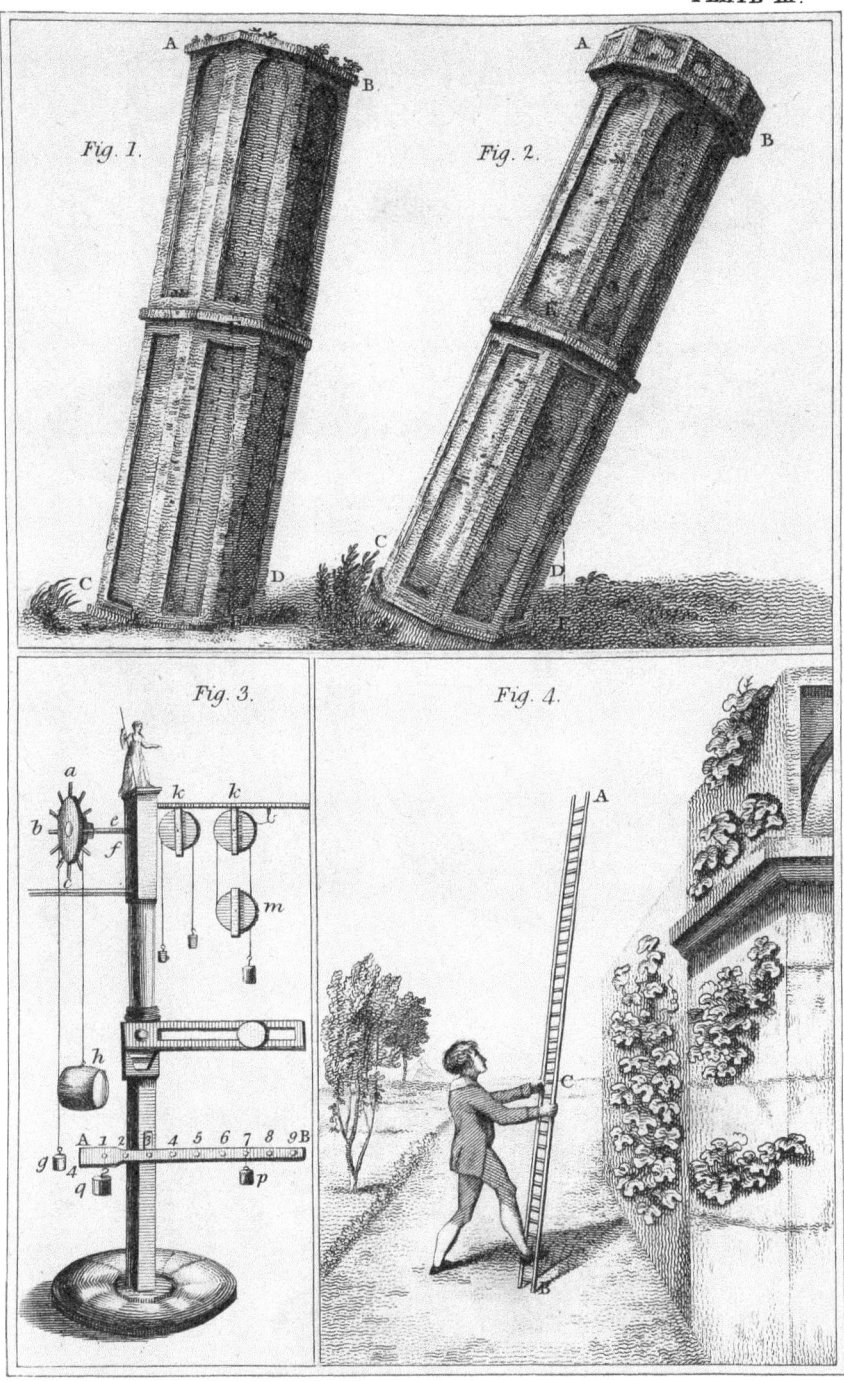

Fig. 1.

Fig. 2.

Fig. 3.

Fig. 4.

M.Bryan design.   T.Noble delin.   H.Mutlow sculp.

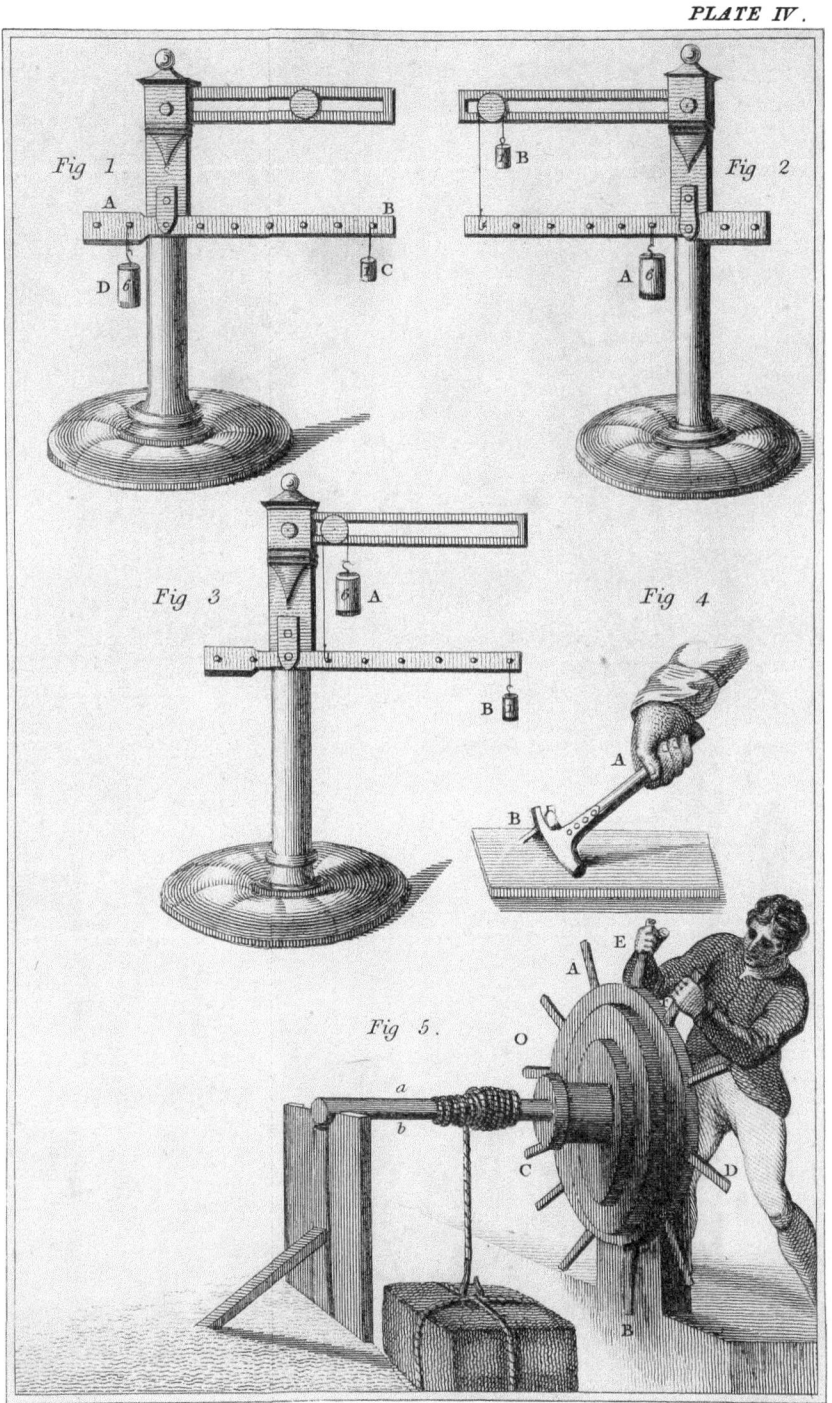

PLATE IV.

Fig 1.

Fig 2.

Fig 3.

Fig 4.

Fig 5.

T. Noble delin.

H. Mutlow sculp.

PLATE V.

Fig 1

Fig. 2

Fig 3

Fig. 4

M. Bryan design. T. Noble delin. H. Mutlow sculp.

PLATE VI.

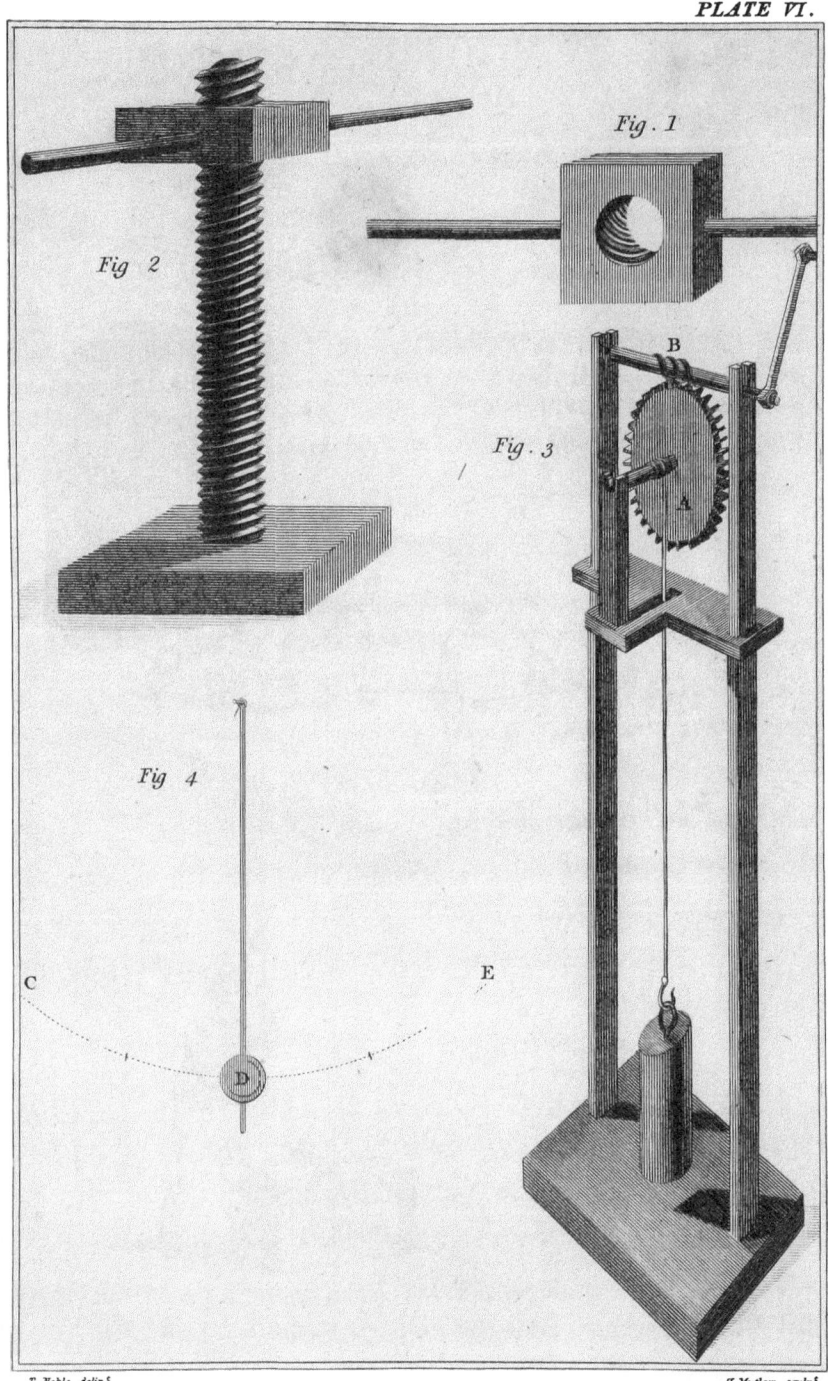

Fig. 1

Fig 2

Fig. 3

B

A

Fig 4

C

D

E

T. Noble delin.ᵗ

H. Mutlow sculp.ᵗ

LECTURE THE SECOND.

MECHANICS.

THE LAWS OF MOTION CONSIDERED.—OF THE CENTRE OF GRAVITY IN
BODIES.—THE SIX MECHANICAL POWERS DESCRIBED, AND THEIR PRO-
PERTIES EXPLAINED.—OF PENDULUMS.—THE EVIDENCES OF THE DEITY
OBSERVABLE IN THE STRUCTURE OF THE HUMAN BODY.—REFLECTIONS.

What cannot Art and Industry perform,
When Science plans the progress of their toil !   BEATTIF.

MATHEMATICAL principles being the foundation of Mechanics,
and all the operations of nature being concerned in the laws of motion; it appears natural in order, to treat of those laws in relation to
mechanics, prior to investigating their effects in the various subjects
of natural philosophy.

The laws of motion may properly be called laws of nature; for
they are rules prescribed to all natural bodies by the great Creator.
In treating of these laws in reference to mechanics, we are to consider four things : A body to be moved—the power which causes the
motion—the space to be passed over by the moving body—and the
duration of the motion. Motion is also considered as simple, or as
compound. Of simple motion three things must be observed; the

velocity, the force, and the space or time of the motion. The degree
of velocity is determined by the space passed over in a given time.
The force or momentum of a body in motion, depends on its velocity, compounded with its mass; and both the velocity and force
of a body in motion, are relatively considered according to the velocity and force of another body in motion. The velocity is a term
which may express the force resident in each equal part of matter.
Suppose one body in motion to contain five times more ponderable
matter than another body in motion: to give each body the same
degree of velocity, we must impel the heavier body with five times
more force than the lighter one. The forces will be equal in two
bodies of equal masses, when they move with the same degree of
velocity; and two bodies of different masses will have equal force,
if the velocity of the smaller body be increased in as great a proportion as its mass is exceeded by the larger one: therefore a small
body may do as much execution as a large one, by having its velocity increased in a due degree. This accounts for the disuse of
the ponderous offensive and defensive machines of the ancients,
since the effect of gunpowder has been discovered; which impels
small bodies with such rapidity, as gives them all the force of those
heavy engines, that from their size and complex structure must
have been very inconvenient and troublesome.

Various are the opinions of historians on the origin of gunpowder. From ancient history we learn something like it was employed in the wars of Alexander, to repel his incursions, by frighten-
ing the elephants, &c. with the inflammable matter thrown among
them. Polydore Virgil ascribes the more modern use of gunpowder
in artillery to a chemist; who, making experiments with the materials
of which it is composed—namely, nitre, sulphur, and charcoal—and
accidentally leaving the composition in a mortar, partly covered with

a stone; a spark fell into it, and threw the stone to a very considerable height: which circumstance excited his attention, and induced a further trial of the composition.—The effects of gunpowder
are certainly dreadful; yet as some medium of offence and defence
has ever been used, I conceive the art of gunnery is not more cruel
than many other destructive devices.

OF THE LAWS OF MOTION.

THE first law relative to natural bodies is, that every substance is
at rest, unless acted upon by some power to cause it to move; and,
on the other hand, a body receiving a given impulse moves in a
straight line with an uniform motion, unless another force acting on
it draw it from that direction, and increase or decrease its degree of
velocity. The second law of motion relates to change of direction
and velocity, which must be in the direction of the impressed force,
and in proportion to it. The third law of motion is, that action and
reaction are always equal and contrary to each other; and therefore,
if two equal forces act on a body in direct opposition to each other,
they destroy all motion. In order to ascertain the difference of
force between two bodies which differ both in mass and velocity,
we must multiply each mass by its respective velocity, and the
difference will appear; as thus: suppose one body contains ten
times as much ponderable matter as another body—say sixty pounds
for the heavier, and six pounds for the lighter one;—also that the
larger body moves with half the velocity of the smaller one; the
velocity is as two to one: $60 \times 1 = 60$; $2 \times 6 = 12$; as $60 \div 12 = 5$:
such is the ratio of their forces, one moving with five times more
force than the other: the velocities are in the exact proportion of
the times. Suppose in one second of time a body moves with a
velocity of 32°, in the next second it will move with 64°, and in

E

the third second with 96° of velocity : but the spaces are in a greater
increasing proportion; for the body which describes 16 feet in one
second of time will describe 48 feet in the next, and 80 feet in the
third second; the spaces increasing according to the odd numbers.
Therefore in the first second of time it will have described 16 feet,
in two seconds 64 feet, in three seconds 144 feet.

The idea of the application of gunpowder to the purposes of fire
arms was first intimated by Roger Bacon, in the thirteenth century.
In 1280 he wrote a treatise on this application of it. His ideas have
since been much improved on; and the theory of projectiles is clearly
understood in the present day, through the effective energetic mind
and liberal communication of that great and most sublime mathe-
matician Dr. Charles Hutton, who has calculated these effects with
the minutest accuracy, and brought this science to the greatest perfection.

Matter contains an innate force to continue its motion in a rectilineal direction, when one power only acts on it; as we may perceive on letting a revolving body suspended from a string go off from
the hand, which proceeds in a rectilineal direction, because only one
force acts on it. While the hand retains the string it is acted upon
by two forces, one called central, the other projectile; which two
forces cause all bodies subjected to them to revolve in a curve.
Every body revolving round a centre endeavours by its vis inertia to
fly off from that centre, which receding force is called centrifugal;
the contrary force, which limits the divergence of the body, is called
centripetal. All bodies moving circularly with great velocity endeavour to fly off from the centre round which they revolve.

The power which causes bodies to persevere in a state of rest is called vis inertia, and is the same as causes them to proceed in straight lines when in motion. As all motion is naturally rectilineal, a body projected in that line would go on for ever in it, if some other power did not urge it in a contrary direction: therefore, when a body describes any other line it must be acted on by two unequal forces. The change of motion in a moving body is in a direction proportioned to the effects of each force impressed: the degree of velocity is also in proportion to the impressed force. Therefore, when two equal forces act on a body at right angles to each other, the body will describe the diagonal of a square, and in the same time that it would describe one side of the square by one force only.

Let a, fig. 1, pl. 2, be supposed a vessel at sea, the wind setting directly from a to b, with such a force as would carry the vessel from a to b in five minutes. Suppose a current of water to set in from a to d with equal force, so that it would of itself carry the vessel from a to d in five minutes; then the vessel will be carried by the joint efforts of those two equal forces, acting at right angles to each other, through the diagonal a e c, of the square a b c d, also in five minutes. In crossing a river in a boat, the current of water frequently acts exactly at right angles to the power or man's strength; when instead of the boat arriving at the directly opposite point, as from b to c, fig. 1, pl. 5, to the place of embarkation, it will describe the diagonal line a e c. A ship, fig. 2, pl. 2, would describe the diagonal of a c, the oblique parallelogram a b c d, fig. 2, pl. 2, if two equal forces acted in oblique angles to each other. Suppose the wind to set against the vessel in the direction from a towards b, and the current of water from a towards d, with equal forces; then the ship would describe the diagonal line a c of the oblique parallelogram a b c d, and in the same time as by one force it would describe A B, or A D. If the two forces are unequal and oblique, the diagonal described will be that of an oblong parallelogram, like A B C D, fig. 3, pl. 2. For, as much as one force exceeds the other, such will be the difference of the length of the sides of the parallelogram. If one force be a continual impulse like the force of gravity, and the other only a force once impressed, then the line described would be a curve *.

The equality of action and reaction is the third law of nature; these being always equal, and in a contrary direction to each other, in all bodies whatever.

Bodies are either elastic or non-elastic. When a body in motion strikes against a body at rest they will both move in the direction of the striking body, but with a velocity in proportion to their quantity of matter. If two bodies strike against each other, moving in the same direction, but with a different velocity, the motion produced will be an intermediate degree between their respective velocities. When two such equal bodies, moving with the same velocity but in contrary directions, impinge against each other, the motion of both will cease ; if moving in contrary directions, and with different velocities, both bodies will move in the direction of that which has the greater force, and the quantity of motion will be in proportion to their difference of force. The inert force in bodies is according to their elasticity; the force a body in motion exerts is in proportion to its velocity multiplied by its weight, as before mentioned; which force is called the momentum of the body.

* For the application of these laws to the heavenly bodies, see my Astronomy.

It is gravity that causes bodies to fall towards the earth with an accelerated motion; yet that acceleration in bodies near the surface of our earth is not ascribed to the attraction of the earth acting stronger on it as it approaches its surface, because the greatest difference is not sufficiently considerable to cause any at the intermediate distances, as it does with the heavenly bodies in their attractions: but the increase of velocity arises from the continuance of the impulse; for, if we impel a body in any direction, and continue the impulse by repeated strokes on it with the same force, the body will go on with an increasing velocity, as bodies do in falling towards the earth. The force causing this acceleration, and occasioning this tendency, is called gravity. This force is always present and acting: therefore bodies approaching the earth by the constant action of this power have their motion uniformly accelerated; and those which are projected from the earth have their motion retarded by it in the same proportion.

We will proceed to consider some other circumstances of heavy bodies necessary to our knowledge of mechanics. There is in every body a point called the centre of gravity, round which point every part is in equilibrium; we may therefore consider the whole weight of the body to be concentrated in this point. To balance a piece of stick on our finger, we support that part which is the centre of its gravity: on moving the finger either way from that point, one end of the stick will preponderate. As the effort of gravity is in the line perpendicular to the horizon, a body will not rest except the centre of gravity and point of suspension are in the same vertical line, which line is called also the line of direction; for, when a body falls towards the earth it is in that direction: a body suspended by a line from a fixed centre, called a plumb-line, rests only when that line is perpendicular to the horizon.

A body placed on a horizontal plane having the line of direction passing within the base will stand, but if that line pass without the base the body will fall: for when the line of direction passes through the base the weight of the body is supported by the surface on which it rests, but when the line of direction passes without the base the body falls, because the centre of gravity loses its support. A B C D, fig. 1, pl. 3, represents an obelisk, which, though it inclines to the horizon, yet having the perpendicular E F from its centre of gravity E within the base, will stand. Let A B C D, fig. 2, pl. 3, represent the same obelisk with an additional gallery, which raises its centre of gravity E so much, that the line of direction E F falls without the base; a body in this situation could not stand. By this we may conceive that when a boat or carriage is likely to be over-set, to prevent that inconvenience, if possible, we should endeavour to sink down so as to bring the line of direction nearer to the base. The broader the base of a body is, and the nearer the line of direction is to its centre, the firmer it will stand; for when the base is narrow and the line of direction near its edge, the body is easily overset. Thus we perceive that the stability of a body on a horizontal plane depends on the position of the line of direction relative to the base of the body. In walking, running, riding, sitting, standing, &c. man uses his balancing powers, without noticing them; yet but for this involuntary motion he could not preserve himself in either of them with grace and security.

The mechanic art is of the greatest antiquity; for in the earliest ages men were compelled to practise it, in order to render the boun-ties of Providence subservient to their necessities. Rude and im-perfect were the first implements employed for the purposes of agri-culture, defence, and conveyance: yet their utility was confirmed by practice; and those subsidiary principles served for a foundation for more modern improvements, the result of greater experience. The great mechanical genius of the ancient Romans, Greeks, Egyptians, &c. is evident, by the vestiges still remaining of their wonderful performances, which, even in their present mutilated state, serve to astonish a modern mechanic. The extensive application and important uses of the mechanic art render it a most noble science: and we are greatly indebted to those who cultivate their genius on a subject of such importance; as thus we learn how to improve every power in nature, and derive every possible advantage from the elements, and are enabled to render them subservient to all the purposes of life.

In treating of mechanical engines, we must consider, first, the weight to be raised; secondly, the power by which it is to be raised; and thirdly, the instrument or engine by which it is to be effected. The number of parts, and the arrangement of them, in mechanical engines, vary according to the effects they are designed to produce; but, however varied and multiplied, we find they are only a combination of six simple powers—the lever, the pulley, the wheel and axis, the inclined plane, the wedge, and the screw; which are constructed to communicate motion to bodies, and to sustain a pressure which a natural power individually could not, perhaps, support: but by dividing it among a certain number of these powers the weight sustained individually is diminished, and the resistance overcome. In the application of the mechanical powers, we are to consider the power and the resistance, and their support and velocities; the centre of gravity, and the line of direction.

A lever is a bar of metal or wood, as represented A B, fig. 3, pl. 3, turning on a prop or centre, commonly called a fulcrum;

the lever is used either to raise weights or overcome resistances. There are three kinds of levers, and in all of them the velocity of each point is directly as its distance from the prop. When this prop is between the weight and the power it is called a lever of the first kind; and as much as the power is placed further from the prop than the weight, or its centre of gravity, so much is the advantage gained by the power over the weight. Scissors are levers of the first kind: the rivet is the centre or prop; the rings are at the longer arms of the lever, where the power is applied; and the points at the extremities of the shorter arms of it, where the resistance is placed. A lever of the second kind is where the weight or resistance is between the prop and the power, as in doors turning on hinges; the hinges are the prop, the door the resistance, and the hand applied to it the power. When the power is applied between the weight and the prop, it is a lever of the third kind; as in removing a ladder from a wall, and supporting it nearly in a perpendicular position. The upper part, A, fig. 4, pl. 3, is the weight; the ground B the fulcrum; and the man the power—which power acts at c, nearer to the fulcrum than the weight does. In the first and second kind of lever, as much as the power acts further from the fulcrum than the weight, so much is the advantage gained, and the power moves so much faster than the resistance; but in the third kind of lever the weight moves faster than the power, and no mechanical advantage is gained by it, but the contrary.

In the wheel and axle, the advantage of the wheel over the axle is as their diameters. If the radius of the wheel *a b c*, fig. 3, pl. 3, be four times as large as the axle *e f*, a power only one-fourth part of the weight of the resistance will support it; as suppose *g* to be a weight of ten pounds, and *h* a cask weighing forty pounds, the former being applied to the circumference of the wheel, and the latter to its axis; for the longer radius of the wheel affords the advantage of a lever the length of itself.

Pulleys are small wheels moved by cords passing over them, and their blocks are either fixed or moveable with the weight. Upper pulleys in fixed blocks, as *k*, fig. 3, pl. 3, serve only to change the direction of the power, giving no mechanical advantage; but under pulleys, as *m*, fig. 3, pl. 3, moving with the weight, give an advantage proportionate to the number of cords by which the weight is sustained. A running or moveable pulley doubles whatever advantage was gained by the other parts of a machine before it was applied; always allowing for friction, which in pulleys is very great.

If machines could be made to move without friction, the least degree of power applied to that which balances the weight would be sufficient to raise it; but as all bodies are full of pores and asperities, these will in some degree retard their motion. The friction in the lever, and in the wheel and axle, is very small: in pulleys it is very considerable; and in the wedge and the screw it is very great indeed.

In the lever we must consider three things:—the prop by which it is supported, and on which it turns; the power that is to raise and support the weight; and the weight to be raised and sustained. A power acts more efficaciously against a weight, in proportion as the weight is more distant from the fulcrum:—two equal masses acting in opposite directions, on the same lever, cannot be in equilibrium but when they are equally distant from the fulcrum:—two unequal weights exert equal forces, when the distance of the lighter weight from the fulcrum, exceeds that of the heavier weight,

F

as much as the heavier exceeds the lighter one ; that is, when their distances from the fulcrum are reciprocally as their masses.

We will now proceed to demonstrate these propositions by experiments. The levers, &c. are balanced before the weights and powers are applied, that the experiments performed by them may be correct, and uninfluenced by the weight of the materials. The studs 1, 2, 3, 4, 5, 6, 7, 8, are all at equal distances on the lever A B, fig. 3, pl. 3, from each other. Fig. 1, pl. 4, represents a lever of the first kind, moveable on an axis, supported by a fulcrum. The shorter arm of this lever is made sufficiently thick to balance the longer one ; therefore the lever, being in equilibrium in itself, may be considered as without weight.

Experiment.

The advantages of a power applied to different parts of the bar, are as the distances of those parts from the fulcrum, compared with the distance of the weight from the same. We will suspend a weight of one ounce on the longer arm, at the sixth hook from the fulcrum, and hang a weight of six ounces on the first hook of the shorter arm ; when the weight of one ounce will balance and support the weight of six ounces.

The weight of one ounce, c, in this experiment is the power ; and that of six ounces, D, the weight or resistance to be supported. A man who, without the aid of any machine, can lift eighty pounds weight, may with a lever, when the resistance is placed at one-sixth part the distance from the fulcrum, where his power or strength is applied, lift four hundred and eighty pounds. By which we perceive, that the weight of a heavier body may be sustained by a lighter one; if the latter have as much more distance from the fulcrum than the former, as it is exceeded by it in weight.

Experiment.

In the second kind of lever, or that in which the weight is between the fulcrum and the power; if the power be placed any number of times further from the fulcrum, than the weight is placed, and a power be applied of the same proportion, say one-fifth the weight of the resistance; the weight A, fig. 2, pl. 4, being placed at the first hook of the lever, and the power B at the fifth hook from the fulcrum, this power will support the weight in the position required, because the force of the power is increased in proportion to its distance from the fulcrum. If we fasten a string to the fifth hook of the long arm of the lever, conveying it over the pulley placed directly above, and suspend a weight of one ounce from the end of the string, we shall perceive the weight of one ounce will balance the weight of six ounces suspended from the first hook, on the same side of the fulcrum.

The cord being over the pulley, the power acts with greater effect, by drawing in a contrary direction to the weight: it will be found easier for a man to raise a weight in this way, than by drawing upwards; as in the latter case he must be above the lever, which would be very inconvenient, and perhaps in some circumstances impracticable.

Fig. 3, pl. 4, represents the third kind of lever, which has the power, A, between the fulcrum and the weight B; in which case the power requisite to support the weight, exceeds it as much, as the distance of the weight from the fulcrum exceeds that of the power.

Experiment.

If we fix one end of the cord to the first hook of the longer arm of the lever, then carry it over the pulley placed directly above it, and suspend a weight of six ounces from the end of the cord, we shall see that it requires all that force to support a weight of one ounce suspended from the sixth hook of the lever; as in this situation the weight to be raised has so much greater force than the power applied. This third kind of lever is never used but from necessity, yet all the motions of the animal body are effected in this way: as in raising the arm, which has the whole weight of the hand to overcome; for the muscles that raise the arm are partly above and partly below the shoulder, or fulcrum, to which they are much nearer than the weight which is to be raised and supported by it; so that when we raise the hand with some pounds weight in it, the force employed is almost incalculable; but the Almighty has endowed the parts of the body with sufficient strength to overcome their apparent disadvantages.

What is called a hammer-lever is similar to the first kind, differing only in name and form, on account of its use and mode of application; for if the handle, A, of a hammer (fig. 4, pl. 4) be four times as long as the iron part, B, which is to draw the nail, the bottom of the hammer resting on the board as a fulcrum, the hand applied at the extremity of the handle is the power, which will raise the nail by the iron part of it with four times less force, than would be required to raise it perpendicularly without this machine; and the hand moves four times further from the nail, by which the velocity of the power is increased in that proportion.

The wheel and axis, o, fig. 5, pl. 4, form a machine which acts something like a lever: it is much used in the mechanic arts, and is presented to us under a great variety of forms.

Experiment.

The power is applied to the circumference of the wheel A B C D, the weight to the circumference of the axis, *a b*, round which a cord winds, having the weight fastened to the end of it. The wheel and axis are fastened together so, that one cannot move without the other. Supposing the power to be a man's strength, applied to the handles on the circumference of the wheel, and that the circumference of the wheel is seven times the circumference of the axis; then the man will raise three hundred and fifty pounds suspended from the axis, with as much ease as he can lift fifty pounds without this machine. If a weight were used to raise the resistance, then would fifty pounds weight, placed round the circumference of the wheel, sustain a burthen of three hundred and fifty pounds: for the weight and power will be in equilibrium when the one is to the other, as the circumference of the wheel is to the circumference of the axis; because the power describes the arc of a large circle, in the time that the weight describes a similar arc of a small one, and consequently the former moves with greater velocity than the latter when the machine is in motion, which velocity is proportionate to the weights in the power and resistance. We may consider the line passing through the diameter of both the wheel and its axis as a straight lever; the extremities of which are in motion, and falling with a velocity proportioned to their distances from the centre round which they revolve; and therefore that the advantage gained, is of the same proportion with that of a straight lever; for the diameter of a circle is known geometrically to be in proportion to its circumference. The thickness of the cords which pass over both a wheel and axis are taken into the calculation of their diameters; i. e. their diameters are added to those of the wheel and axis.

A small wheel turning about its axis, with a cord passing round it, is called a pulley As before observed, there are two kinds of pulleys: the fixed, which do not move out of their place, as A, fig. 1, pl. 5: and the moveable, which rise and fall with the weight, as B. A fixed pulley turns on its axis, but does not move out of its place: it merely changes the direction of the power, giving it no mechanical advantage ; therefore no greater weight can be raised by this pulley than by a man's natural strength. Yet it is convenient by altering the direction, and affording the joint aid of several persons to raise a great weight, by each pulling the rope at the same time. The moveable pulley rises and falls with the weight, and thus adds to the efficacy of the force applied. In order to understand the advantage gained by moveable pulleys, we must first consider that each pulley hangs by two cords equally stretched, and which must consequently bear equal parts of the weight; and that when a weight is supported by a cord passing round several fixed and moveable pulleys, as all the parts of it on each side are equally stretched, the whole weight must be equally divided among them: therefore, if the power which hangs at the end of the cord be equal to the divided weight sustained by each division of it, that power must sustain the weight; as the power and weight must balance each other, when the power is in proportion to the weight, as one is to the whole number of parts into which the cord is divided.

A, fig. 1, pl. 5, represents a fixed block containing one pulley, which only turns upon its axis; and B, a moveable block, containing also one pulley, which likewise turns on its axis, and rises with its block and weight: the advantage gained by the latter is two to one. If a system consist of three moveable pulleys with six strings, the advantage gained is six to one; which may be increased in any proportion, according to the number of strings and moveable pulleys.

The inclined plane may with propriety be reckoned among the mechanical powers, as it affords equal advantages under certain circumstances; for it raises weights by diminishing their force when laid on it. It is chiefly used to remove great weights to a perpendicular height, or to let them down into a lower situation: in both cases it diminishes the effects of gravity, and thereby facilitates the motion.

Suppose fig. 2, pl. 5, to represent a barrel of liquor, which is to be let down into a cellar of considerable depth; it would require the united strength of several men to lower it safely without staving the vessel: but if a plank be laid along the stairs, which is an inclined plane, perhaps one man will be able to let it down securely; and here the inclined plane is the only mechanical power that can be readily applied. Here *a b*, are two props, round which the rope coils in order to regulate the motion: the advantage gained, is in the same proportion as the length of the plane exceeds its height. A plane A, fig. 3, pl. 5, that is three times its perpendicular height C D, affords an advantage of three to one; or if the length be four times its height, the advantage gained will be four to one: so that what would require the strength of four men to let down or raise up perpendicularly, may be effected by one man, by the assistance of an inclined plane, the length of which is four times that of its perpendicular height. Suppose C B, fig. 3, pl. 5, to be a bank, to the top of which a collection of plants is to be conveyed from the lower ground, A. For this purpose the labourer uses an inclined plane, up which he ascends pushing the loaded barrow before him, effecting his business with ease and convenience.

The wedge is a mechanical power, employed principally to cleave wood; its form is broad at the back, terminating in a thin edge, as shown fig. 4, pl. 5; the back part of the wedge is struck upon by a hammer, and the percussion, by causing it to enter the wood with force, separates the parts of it with greater ease than by any other machine we use. The hatchet, spade, chisel, and needles, and all instruments which, beginning from edges or points, gradually thicken, may be denominated wedges; at least their effects are the same. Wedges are also of considerable use in raising, and afterwards supporting, great weights; and so much force may be applied in this way, that some thousand tons may be raised by this power, and supported in the situation required.

The screw is a mechanic power, of great use in pressing bodies closely together, and keeping them in that situation; also for raising weights.

The screw consists of two distinct parts, the outside and inside: the outside or receiving part, fig. 1, pl. 6, is cut in such a manner as to have a groove going round it in a spiral form. The inside or propelling part is a solid body; the convex surface of which is cut in the same manner as the concave surface of the outside screw, thus forming a spiral thread round it, so that the former fits into the latter: one part is fixed while the other turns round; and in each revolution the moveable part is carried, in the direction of the cylinder, through a space equal in length to the interval between the two contiguous threads of the screw by which that body moved, passes a space equal to that interval.

When a handle or lever is employed to assist in turning a screw, as much as the largest diameter of the handle is more than the intervals between the spirals, so much is the advantage gained by the moving power. The friction in screws is very great; which is an advantage, as it enables the screw to retain the position it has gained, which the other mechanical powers (excepting the wedge) are not adapted to do. Very great weights may therefore be raised and supported by this power. Though the handle of a screw adds to the intensity of the force, yet the screw could be used without it if the force were sufficient; but the advantage gained is in the same proportion as the length of the handle exceeds the distance between the threads of the screw. A screw increases the power in the same way that an inclined plane does; the screw being generated from an inclined plane. If we cut a piece of paper like an inclined plane, and coil it round a cylinder, it will represent the threads of a screw, that would afford an advantage in the proportion as the hypothenuse is to the perpendicular of it.

When a screw is applied to a wheel, it is called an endless screw; for, turning on an axis, and the teeth of it fitting those on the circumference of the wheel, it may be turned continually. The endless or perpetual screw is represented, fig. 3, pl. 6: this screw is fastened to a handle, which turns it and acts as a lever. A is an indented wheel, which being turned by the screw, its teeth catch the thread of the screw B; and while the screw turns once round, the wheel raises a weight suspended from it through a space equal to the distance between two teeth. The advantage gained by the power

G

or handle, is in proportion as the length of the handle is to the distances between the teeth of the wheel. When a slow motion is required, as in raising a great weight, the endless screw is perfectly adapted to that purpose; for the motion of the wheel may be very slow, and the power of the machine very great.

From the six simple machines I have been describing compounded ones are formed, to serve for different purposes; but, however varied or multiplied, the laws are the same as have been already explained, namely, the advantage gained by the power employed is in proportion to its velocity, or its distance from the fulcrum, compared with the velocity of the resistance, or the distance of that from the same. We readily perceive that with any one of the mechanical powers properly applied, we may overcome every possible resistance; yet it is not always convenient to use them separately of a size sufficient to effect our purposes; therefore by combination we increase our powers, and render the application of them more easy and convenient.

The variety of compound engines precludes a general description of them in this work, and I do not know how to select any that would render the subject more interesting; I shall therefore in this place make only a few remarks of general application in compound engines. In a machine constructed with more than one of the mechanical powers, whatever be the number applied, the advantage gained by their joint action over the weight, will be in proportion to what each power dependant on another power gains: as thus. Suppose a machine to consist of a wheel and axle, assisted by three pulleys having six cords. Let the diameters of the wheel and axis be in such a proportion to each other, that the advantage gained by the power of the wheel over the axis shall be also six times; then shall the two powers be estimated thus, $6 \times 6 = 36$: for by the wheel and axis the advantage gained will be 6 to 1; and the power of the pulleys being also 6 to 1, those numbers multiplied together make 36; therefore a power or weight equal to $\frac{1}{36}$ part of the weight to be raised, will balance it, by the assistance of a wheel and axle the diameters of which differ six times, and a moveable pulley of six cords.

The chief art in compounding machines, in order to save labour and expence, consists in their simplicity, and having as few moving parts as possible.

A pendulum is so intimately connected with the laws of mechanics, and so essential in measuring time, that it seems to require notice in this place, though it is not called a mechanical power.

Fig. 4, pl. 6, represents a heavy body, suspended by a line from its fulcrum or support. On giving an impulse to the thread, B will ascend, suppose to c, in the circle c d c. When it arrives at c, it will fall by its gravity to d, whence it would go on in a straight line by the force it has acquired in falling, were it not prevented by the thread, which changes its direction; but it proceeds by its centrifugal force to the same distance on the other side of a d, to c, that it arrived at by the first impulse it received.

If the motion of a pendulum were not retarded by the friction of the material on which it revolves, or the resistance of the air, the arc it describes in each vibration would be exactly equal, and the pendulum would vibrate for ever. But this is not the case; therefore the arcs described by a pendulum in its vibrations, become less and less continually, till the impressed force which produced the motion is totally lost.

For the uses of the pendulum we are chiefly indebted to Galileo, who, observing the vibrations of a lamp suspended from the roof of a house, discovered that they were performed sensibly in the same time, whether great or small; but that their extent continually diminished till the motion entirely ceased; and that the vibrations were slower as the lamp was at a greater distance from the point of suspension.

Pendulums of the same length vibrate in the same time, whatever be the proportion of their weights. If one pendulum be four times as long as another, the shorter will perform two vibrations in the same time that the longer performs one; for it is the length of the pendulum rod that determines the time of each oscillation.

There is one point within every pendulum, into which if all the matter that composes it were condensed, the time of its vibrations would not be altered; this point is called the centre of oscillation. The point round which the pendulum vibrates, is called its centre of motion.

Huygens was the first who demonstrated the properties of pendulums: he asserted, that if the point to which a pendulum is suspended were perfectly fixed, and all manner of friction and resistance removed, a pendulum once set in motion would for ever continue to vibrate, and without diminution of force. But as this is not the case in practice, a pendulum which has received a certain impulse will, unless that impulse be continually renewed (as in the swing-wheel of a clock), describe less and less arcs every time, till its motion ceases.

The time of each vibration of a pendulum is in proportion to the square root of its length; that of ten inches long performing its vibrations in half the time in which one of forty inches performs its vibrations. Thus, the shorter the pendulum the quicker the performance, according to the above ratio; for if a pendulum oscillates always the same arc of a circle, the longer pendulum must be longer in describing the arc of its circle, that circle being larger.

In order to render a pendulum a measurer of time, a particular machine has been constructed to regulate and continue its motion. The length of a pendulum rod is variable; for heat dilates and cold contracts all metals. Various contrivances have been employed to associate different materials, and combine them in such a form as should counteract the inconvenience of these fluctuations, so as to render the motion of a pendulum equable in all climates and seasons; but perfect accuracy has not yet been obtained in this grand desideratum.

It appears evident, from what we learn of the direction of power in the limbs of men and quadrupeds, that the former were designed to move upright, and that they can bear a burthen better in that position than in any other. Two men with a burthen between them, will carry a greater weight, than double what each can separately; because by using a pole, they can preserve such a position that the whole pillar of their bones supports the weight.

If one man be twice as strong as the other, the weight should be moved towards the stronger man in that proportion, namely, to half the distance from the latter that it is from the former; by which means the weaker man will bear only one-third of the burthen. In attaching a pair of horses to a carriage, if one be weaker than the other, the stronger horse should be placed nearer the centre of the beam, that is fastened to the carriage, than the weaker, by which means each will draw in proportion to its strength; the motion of the carriage will also be facilitated by this equipoise of power.

We may infer from the direction of power in horses, that they were designed to draw burthens; and may suppose that a horse will draw a weight in proportion to his strength: yet it is easy to understand, from our observations on action and reaction, that two horses of unequal strength may draw the same weight; or the weaker horse may even draw a weight the stronger cannot remove, if the weaker be the heavier, or exceed the other more in weight than he is exceeded by him in strength; for a weight reacts and pulls back a horse in proportion to itself: therefore the heavier horse, though he be the weaker, will, if his weight be greater than the strength of his antagonist, lose less power than the stronger one. A horse has two sources of power in drawing weights: his strength, which gives him velocity: and his weight, which gives him force. Horses must have sufficient force or weight to enable them to move a heavy carriage; for if they have not, they cannot secure their feet on the ground, but will slip, and be drawn backwards.

In ancient times sledges were used to convey heavy goods; but since the power of wheels has been ascertained, they have been adopted, as better calculated to overcome obstacles and diminish weight: for a wheel turning on an axis, diminishes the resistance, and facilitates the motion, and thereby renders the draught easy ; but a sledge passing over a road, meets with many obstructions to retard its progress, and diminish its velocity. All wheels act as levers, and their power is in the proportion that their whole diameter exceeds that of their axis.

The nearer the centre of gravity of a carriage is to the ground, the safer it is; and to render the draught easy, the centre of gravity should be as near as possible to the axis of the wheel or centre of motion. Springs, by their elasticity, facilitate the motion of carriages; for they allow the wheels to pass gradually over obstacles, and diminish the effects of gravity.

Though the friction of sledges is very considerable on uneven roads, yet in Lapland and other cold countries, where the surface is frozen snow or ice, no other vehicle can be used either with safety or convenience ; for wheels would cut the ice and sink into it, nor would they readily turn on such a surface. Wheels are kept turning on our roads by the obstacles they meet with, which for a moment arrest their motion, and occasion a suitable reaction. The resistance is not felt on the circumference of the wheel, but at the axis, which is not more than a few inches diameter: the force acts on the circumference of the wheel, which being at a much greater distance from the fulcrum or axis, the advantage of the wheel, is in proportion to its diameter compared with that of the axis; so that large wheels have the advantage over small ones, in facilitating the progress of a carriage. It is necessary, for the convenience of turning in a shorter space, that the front wheels of carriages should be smaller than the hind ones, that the carriage may pass over them ; but otherwise carriages would move lighter were both pairs of wheels of the larger size. In loading waggons, the greatest weight should be laid behind; but in loading small carriages with one pair of wheels, if the load be laid too much behind in going up hill, there is danger of lifting up the shaft horse, and so overturning the vehicle. Broad wheels are made very convex, to support the load when they fall into ruts; as by this contrivance the weight rests perpendicularly, not obliquely, on the wheels.

OF MAN AS A MACHINE.

Let us turn from the mechanism effected by the art of man, to contemplate that of his Maker: the one feeble, and easily estimated; the other strong, and incalculable.

Behold that various and complicated machinery, which forms the graceful column of man! composed of bones, joints, and arteries; and clothed with muscles, veins, and teguments! How duly balanced! How aptly contrived for his various movements! As the summit of this column, the head appears, appointed to this highest station as containing the seat of sensation, the light of understanding, and the faculty of sight. To effect its most extensive purposes, it moves on an articulated fulcrum or prop, on which it can turn either backward or forward, up or down, horizontally to the right or to the left. The first two movements are effected by a hinge-joint fitted to the upper vertebra of the neck bone, but limited by ligaments, in its movements backward and forward, to prevent suffocation. The horizontal motion is effected by a particular auxiliary, placed on the bone below the first vertebra. It is a prong, or projection, similar in shape and size to a tooth, which fits into a pivot of the bone above it, and serves as an axle for the head to turn, but only to a limited distance, muscles on each side guarding it from excessive danger. The utility of the head moving on a

hinge on the first vertebra, instead of the vertebra having the vibratory motion, has been discovered by anatomists; namely, that
had that vertebra this motion, the spinal marrow at the beginning
of its course would be impeded. The spine or backbone is a continuation of vertebræ: its various uses, and amazing powers, can
never be wholly investigated; yet we can perceive enough of the
design to compel us to adore and venerate the superior excellence
of the work of God! Various and almost contrary are its properties,
yet all are applied in the most efficacious manner possible. It is
firm enough to bear the body in an erect posture, flexible enough
to allow of various curvatures to relieve the body when fatigued: it
connects the spinal marrow with the brain: it is the grand conduit
or pipe, which, by its various mouths, conveys the matter of the brain
to various parts of the body, through innumerable capillary tubes
called nerves. This pipe affords support to all the muscles of the
trunk of the body, which are attached to it: the ribs also are articulated into the vertebræ of the back. Were an artist to attempt to
form a chain which should be capable of supporting a weight perpendicularly, and at the same time be flexible, he would make it
of strong and short links, and endeavour to combine flexibility and
strength, so as to act in opposition to each other; but he would find
it very difficult to effect his purpose, even for supporting a small
weight in this manner.

There are twenty-four bones in the human spine, joined to each
other by broad bases; in some parts these bases are shallower than
in others, according as they are to serve more immediately either the
purposes of flexibility or strength. In the back, where strength is
most wanted, they are firmer than in the loins, where flexibility is
necessary; and still firmer in the neck, where the erect posture is
chiefly required. Each of these bones is perforated through the

middle, and so placed over and under those next to it, as to form a close channel for the medullary substance. To prevent this channel from being disturbed on change of posture, by the vertebræ shifting over one another, these bones are supplied with cartilages, which, being of an elastic and yielding nature, allow of these motions without any separation of the bones themselves.

Having briefly considered the fulcrum or principal prop of that great compounded machine the human frame, with all its movements, we may proceed to consider some of its component parts separately, in their mechanical construction and various forms; and observe how admirably they are adapted to effect their different purposes; in all of which we shall perceive the wonderful excellence of the work of our Creator: though enriched by the learning of ages, the most illuminated genius cannot fully comprehend its incommensurate wisdom and utility.

OF THE JOINTS.

ON the various joints of the bones much of their different effects depend. Each is mechanical, and resolvable by human reason. There are two sorts of joints; namely, the ball and socket, and the hinge joint; and one or the other is used according to the motion required. At the knee, a hinge answers the purpose of moving the leg backwards and forwards: at the hip, a ball and socket serve to co-operate with the motion of the leg, and also to move the limb to the right and left in any required position. The shoulder joint is likewise a ball and socket, but the socket is shallow and has a cartilage set round its rim; while the cup of the thigh bone is very concave, and made of more solid materials. These differences agree with the situations of each of them, and the purposes they are separately to answer; for, as the one is a principal instrument of action, the shallowness of the socket, and flexibility of the cartilage, form its motion : while in the leg and thigh, which are to support the body, firmness is likewise necessary, which has been considered in the conformation and texture of the joints belonging to each of them. In all the joints of the body the ends of the bones are covered with gristle, to prevent injury by the friction of hard substances. The ball is tipped, and the cup lined, with this yielding substance; and the hip joint is protected by it. Each hinge and socket is also supplied with a mucilaginous ointment, which constantly softens and lubricates the parts. This juice, called synovia, is supplied by glands, so placed that on each motion of the joint, it is pressed out as from a sponge.

Having considered the joints on which animal motions depend, we may contemplate the mechanism by which these motions are generated and supported. The muscles and their tendons are not only constitutionally endowed to generate and regulate motion, but also differently constructed for these purposes, according to the movement required, and the instrument used. Thus, at the elbow and knee, where there is a hinge joint, which serves only to move the limb in the same plane, the leaders, or muscular tendons, are placed parallel to them, and lengthen and shorten only in that direction; but in the shoulder and hip, where the ball and socket joint is found, the muscles are variously placed, and are capable of contracting and restoring themselves in each position. The muscles also, by their different directions, support the bones, particularly the head; and all the limbs are regulated in their movements chiefly by their agency. Each muscle has what is called an antagonist muscle; namely, one that acts in a direction contrary to the other. For the muscles cannot expand beyond their natural state, though they can contract; therefore to produce a contrary motion, a new muscle must be called into action. It is by this contrary motion of the muscles of the face, that the features are duly balanced in their places. The natural strength of the muscles may be either increased or diminished by exercise; for we perceive the legs of a porter, and the arms of an anchorsmith, are stronger by use. All the limbs of the body are levers of the third kind; for the resistance must be further from the prop than the power, the power being in the joint itself: yet all is easy, no difficulty arises—for it is the work of God!

How do the limited and obscure effects of men's understanding sink before the unbounded, simple, and effective energies of the Divine mind! We have been contemplating them in familiar subjects; but let us extend our intellectual view to the consideration of all created bodies within our sphere of knowledge, though we cannot conceive their profundity, or penetrate their essence. How stupendous the Wisdom that created all things!—that appointed each particle of matter its appropriate office and station, velocity and direction!—endowed the larger bodies of this beautiful universe with active properties and influences, prescribing them certain limits, certain beneficial boundaries, which they can never exceed nor disobey!—that contrived the curious texture of plants and vegetables, as best adapted to their growth and nourishment, and the organs of men and animals for their respective functions! How does finite wisdom and execution diminish in our sight, when compared with the infinite excellence of the works of the great Creator! The weak efforts, the faint effects of men's understanding, fade before infinite Wisdom and Performance, like the taper's faint and glimmering light before the sun's exalted, pure, and effulgent radiance.

Man at home, within himself, can find
The Deity immense ; and in that frame,
So fearfully, so wonderfully made,
See and adore his providence and power.
I see, and I adore.—O God ! most bounteous !
O infinite of goodness and glory !
The knee that thou hast shaped shall bend to Thee ;
The tongue which thou hast tuned shall chaunt thy praise :
And thine own image, the immortal soul,
Shall consecrate herself to Thee for ever.   SMART.

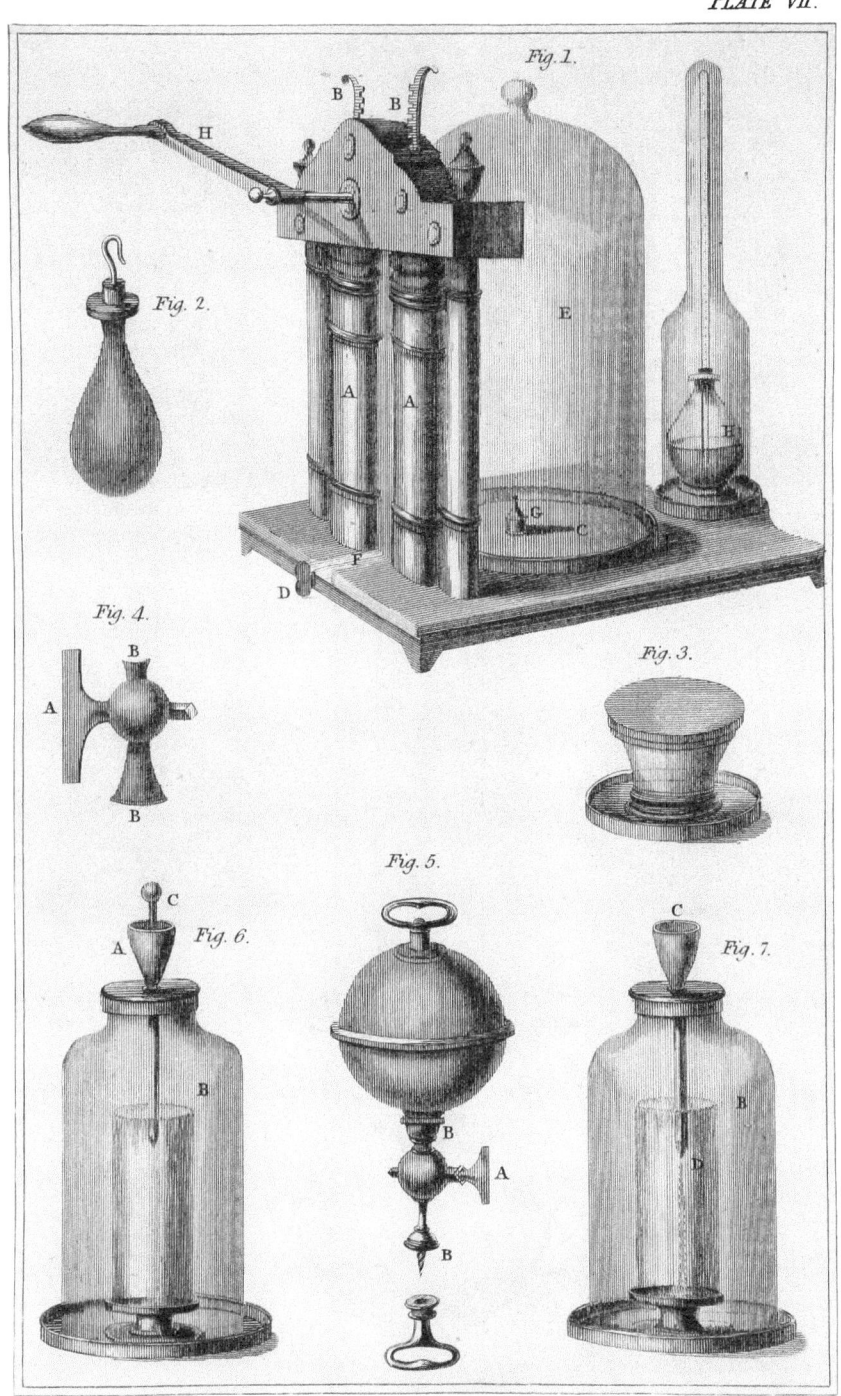

PLATE VII.

Fig. 1.

Fig. 2.

Fig. 4.

Fig. 3.

Fig. 5.

Fig. 6.

Fig. 7.

T. Noble delt.

Mutlow Sculp.

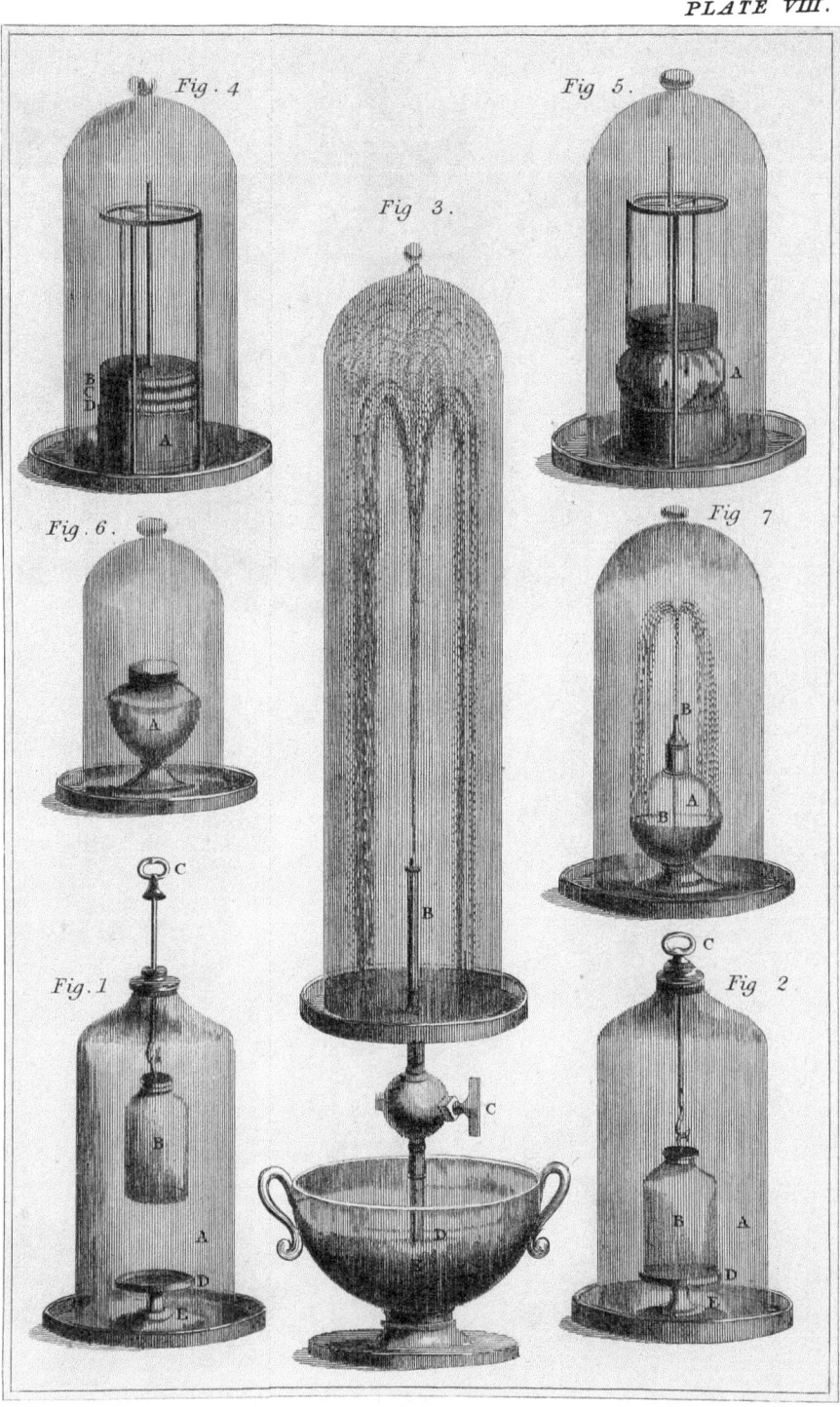

PLATE VIII.

Fig. 4

Fig 5.

Fig 3.

Fig. 6.

Fig 7

Fig. 1

Fig 2

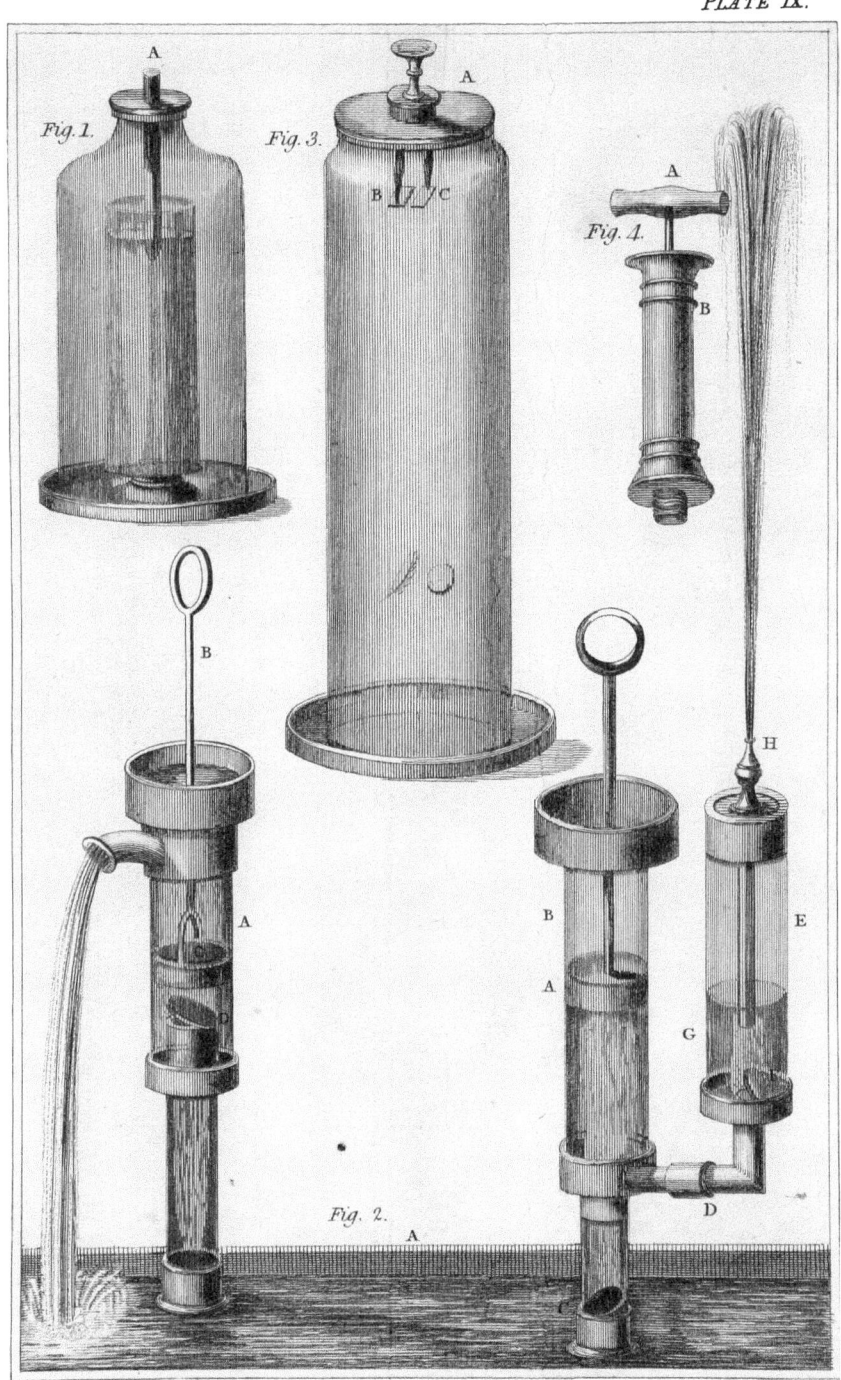

PLATE IX.

Fig. 1.

Fig. 3.

Fig. 4.

Fig. 2.

T. Noble delt.

Mutlow Sculpt.

PLATE X.

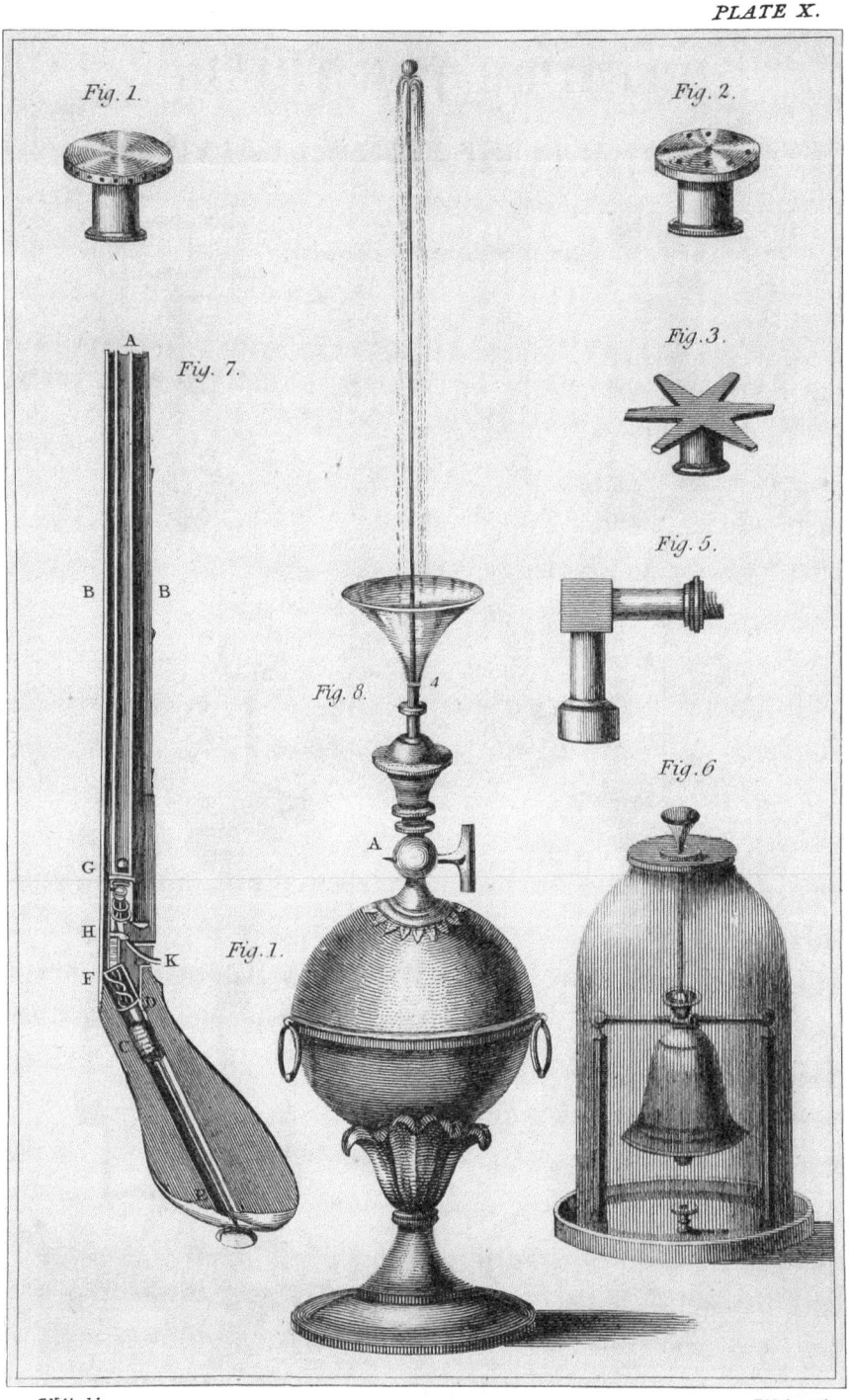

Fig. 1. Fig. 2. Fig. 3. Fig. 5. Fig. 6. Fig. 7. Fig. 8. Fig. 1.

T. Noble del. H. Mutlow sculp.

LECTURE THE THIRD,

AND THE FIRST ON PNEUMATICS.

OF THE AMAZING WEIGHT OF THE ATMOSPHERE, THE ELASTICITY, AND THE OTHER NATURAL PROPERTIES OF THE AIR, EXHIBITED BY A VARIETY OF PLEASING AND INTERESTING EXPERIMENTS.

God made
The firmament, expanse of liquid, pure,
Transparent, elemental air, diffused
In circuit to the uttermost convex
Of this great round.   MILTON.

HAVING contemplated all that is necessary for comprehending the elementary characteristics of earth and fire, we may proceed to consider the nature and properties of air; a subject so curious, that it will astonish those who have not previously observed the wonderful and great effects of this invisible element.

Previously to performing the experiments on air, it will be necessary to understand the construction and operation of the machine used to exhibit them. The invention of the air-pump is ascribed by some persons to Mr. Boyle: but according to others, the discovery was not made by that amiable philosopher; the original idea belonging to Otto Guericke, the celebrated consul of

Magdeburg, who exhibited his experiments with it before the empire and states of Germany, at the breaking up of the imperial diet at Ratisbon, in 1654. However, this machine was certainly much improved by the great Boyle; who, assisted by Dr. Hook, brought it to a considerable degree of perfection: since his time, it has been rendered still more perfect; so that what was originally performed with great labour, is now effected in the easiest manner possible.

DESCRIPTION OF THE AIR-PUMP.

Fig. 1, pl. 7, represents an air-pump. a a, the barrels; b b, the piston-rods, moveable in them; c, the plate of the pump, on which the receivers are placed, where the air is extracted from them; d, a screw to readmit the air when necessary. The barrels are cylindrical, and smooth. The piston-rods have each a piece of metal at the bottom, exactly fitting the aperture of the cylinders. When down, they lie close over valves placed at the bottom of the barrels: these valves allow the air to pass upward out of the top of the barrel, but prevent its return. e is a glass receiver, under which is placed any thing from which we wish to extract the air. Receivers should be very accurately ground on their edge, that they may lie close to the plate of the air-pump, to prevent any air rushing into them; but, in general, it is necessary to have an oiled leather previously laid on the plate of the pump, to fill up any little interstices, and effectually prevent the ingress of air. f is the canal for the air to pass from under the receiver to the barrels with which it communicates; g is a tube forming the communication between the receiver and the barrel. Air-pumps are variously constructed, but this explanation will sufficiently serve to explain the nature and operation of all machines used for the same purpose. To work this pump, we turn the handle H, which raises one piston-rod and depresses the other. The one raised lifts up the whole column of air that was within the barrel; which column cannot return, as the valve does not open downward; but the air from under the receiver will rush into the barrel, through the valve at the bottom of it, and restore the equilibrium. Every thing by us denominated empty is, naturally, full of air. This air is a substance of an active and elastic nature; and is always endeavouring to place itself in equilibrium; therefore, whenever it is disturbed, it immediately exerts itself to restore the balance. By the operation of the air-pump, the quantity of air under the receiver is diminished, and the residue occupies a larger space. By continuing the process, we are able entirely to exhaust the receiver, for by alternately raising the piston-rods, we lift up a column of air with every stroke.

I cannot enter on a description of the surprising energies of air, without previously calling your attention to that consideration of it, which will make the experiments I shall have the pleasure of showing you, not only gratifying to curiosity, but tributary to a better purpose.

The facility with which we breathe, prevents our considering the power that enables us to do so; yet, as living and breathing are synonymous, surely that which is the medium of our existence deserves our attention, and must be an object worthy the investigation of rational creatures.

The globe which we inhabit is surrounded by a medium we call atmospheric fluid; this is composed of air and a heterogeneous collection of particles raised from all bodies on its surface, by effluvia, exhalations, &c. The learned have considered it as a large chemical

vessel, containing the matter of all sublunary bodies; and, in consequence of being exposed to the amazing heat of the sun, producing the various operations of sublimation, separation, composition, fermentation, &c. The electric fluid is a material element
in this compounded mass; it pervades all parts of it, and from its
influence principally arise meteors, hail, &c.

While the experiments on air serve for amusement, by displaying the affections of the atmosphere, the usefulness and indispensability of this fluid medium render the consideration of it a
matter of the highest importance to us; for we perceive that without the air no animal could exist, no vegetable could grow, no
animated form could enliven and embellish this beautiful world.

We will now proceed to examine the wonderful properties of
this medium of all existence. We do not perceive the impressions
of air arising from its weight; that affection of the atmosphere being
counteracted by the wisdom of Providence, which has supplied the
internal parts of our bodies with this fluid to counterbalance its external force. On extracting the air from the receiver, the pressure
of the atmosphere will be evident to us, and we shall adore that
Beneficence which prevents the effects of that pressure on our bodies
which would certainly annihilate them.

Experiments on Atmospheric Pressure.

THE vessel E is full of air; I will place it on the plate of the
pump, whence we can remove it with ease. By working the machine we shall exhaust the receiver of its air, when it will be found
impossible to remove the receiver from the plate by any force we
can exert. It is the weight of the external air on the surface of the
glass that keeps it down; the air that was within it, or the counter-

balance, being removed. We could never lift the glass from the pump without breaking it were it not for the screw D, which, on being turned, suffers the external air to re-enter the receiver, and restore the equilibrium: this performed, we can remove the receiver as easily as before, because the counterpressure is restored.

This experiment is sufficient to convince us that the atmosphere has weight; yet, as a further evidence of this truth, we will weigh the flask, fig. 2, pl. 7, when full of air, and also when the air is exhausted from it: the weight of this small quantity of air is but trifling, compared with the weight of a whole column of the atmosphere.

Every square inch of the exterior of our bodies, and on the surface of the earth, supports nearly fifteen pounds weight of the atmosphere; so that a middle-sized person sustains a pressure of air equal to 30,240 pounds, estimating his surface at fourteen square feet, which we may suppose it to be at a mean calculation.

Air, being a fluid, gravitates in all directions: of which we are convinced by its pressure horizontally into the canal of the pump, and also against the inside of the receiver; which is in every direction, and likewise equal to the compressing force of the air on its external surface. Thus, we do not feel the external pressure of the atmosphere on our bodies, because it is within us, and the internal resistance is endowed with a power equal to the compressing force. We may readily conceive these effects of the external and internal air to be necessary to animal existence, in order to keep up a constant motion in the fluids of our bodies; and that if our bodies did not contain air, the external air would press us to death; or if the external pressure were removed, the internal expansion would burst all the vessels of our bodies, and animal existence would be impossible. Delightful evidences of the wisdom and preventing goodness of our great Creator! Let us never suffer such consoling and invigorating testimonies to escape our due observation and gratitude.

All that is necessary to convince you of the weight of the atmosphere has been advanced; yet, as amusement serves to enliven instruction, we will proceed with a variety of pleasing experiments on this subject.

Experiment.

I have tied a bladder over the top of the open receiver, fig. 3, pl. 7. Putting this on the plate of the pump, and extracting the air from under it, the external air falls so heavily on the bladder, as to burst it in pieces by the force exerted; and a loud report is heard, occasioned by the blow the bladder receives from the atmosphere.

Before shewing you the experiments with a stop-cock, I will explain its construction. A stop-cock acts something like a valve; it is represented by fig. 4. The handle A, when turned upwards, leaves a canal open through the barrel B B; but when A is turned horizontally, it shuts the passage.

Experiment.

The hemispheres, fig. 5, pl. 7, are fixed to a stop-cock. We screw the bottom to the plate of the pump, and turn the handle A upright, to form a communication with the barrels by means of the canal B B. We can now separate the hemispheres with ease; but when the air is extracted from within them, it will be found impossible to do so. We turn the handle horizontally, to prevent the external air from rushing into the hemispheres, when we unscrew them from the plate of the pump; on turning the handle upright, to readmit the air, we shall readily separate them.

Experiment.

I will place the wooden cup A, fig. 6, pl. 7, on the open top of the receiver B, fig. 6, pl. 7: this cup has a long piece of close-grained boxwood affixed to it, and a plug c; which latter is left in at the beginning of the operation. We will pour quicksilver into the cup, and exhaust the air from the receiver; when, on with-drawing the plug, as at c, fig. 7, pl. 7, we shall see the quicksilver descend through the pores of the wood, and fall in a shower into the glass vessel D placed to receive it. This effect is produced by the weight of the external air on the surface of the quicksilver, which forces it through the pores of the wood. We perceive, by this ex-periment, that the course of the air-vessels in wood is longitudinal, for the fluid does not pass horizontally through them. Science is liable to misconception, and there are always persons to raise objections: this has been the case in respect to the most clear demonstrations of the weight of the atmosphere; for it has been supposed, by the unlearned, that receivers might be kept down to the plate of the air-pump by another power, which they, erroneously, denominate suction, without considering the meaning of the term; for suction certainly produces a vacuum, which by no means con-tradicts the fact of external pressure: but by that term they under-stand very differently, for they suppose that the glass is kept down by the moisture of the leather; thereby ascribing to an oiled leather a power it never can have. This absurd assertion hardly deserves notice; yet, to prevent your being diverted from the truth by such vague ideas, I will convince you that the receiver is kept down by no power but the pressure of the external air on its surface, on the internal air being removed.

Experiment.

I will suspend the small receiver B, fig. 1, pl. 8, within-side the large open one A ; and then exhaust the air from both receivers; placing B, by means of the sliding wire c, on the stand D, as exhibited by fig. 2, pl. 8. In this state of the apparatus, by readmitting air to the large receiver A, fig. 2, but not to the small one B, fig. 2, the latter will be so firmly fixed to the brass stand, that we shall not be able to remove it; the weight of the external air keeping it down with great force.

In order to introduce the air under the small receiver, unscrew the stem E of the stand, and the air will rush in through a small hole in the brass plate, and restore the equilibrium. Nothing but the weight of the external air, and the balancing power being removed, could possibly keep the receiver close to the plate of the stand.

I will show you that the elasticity of the air is equal to its compressing force, by repeating the experiment; when, after admitting the air to the large receiver, we shall find the elasticity of the air included in it will keep the small receiver fixed to the stand.

Experiment.

Fig. 3, pl. 8, is called a transferrer. It has a capillary tube, B, fixed to a stop-cock. The stop-cock being screwed into the plate of the pump, I shall place over the tube B a tall receiver; turning the handle upright, to form a communication between it and the pump, on exhausting the air from the receiver: then turning the handle c horizontally, and unscrewing the apparatus from the pump, I shall place it under the surface of the water, as represented at D, in the bason E; when the pressure of the external air on the surface of the water will force it through the pipe, and form a beautiful fountain.

Experiment.

The resistance confined air makes to an external force is evident, by simply pressing a bladder containing a small quantity, when the resistance of the internal air will be felt.

Experiments on the Air's Resistance.

Take a wine-glass, and fill it with water; put a piece of paper over the top, and then place your hand on it: when, on inverting the glass, and removing your hand, the water will remain supported by the air; which cannot act on it in any other direction than that opposite to the weight of the water. That it is the pressure of the air which supports the water is evident by the concavity of the paper.

The knowledge acquired of the pressure of the air has produced a very useful instrument, which indicates the state of the atmosphere for every time and every place. This instrument, called a barometer, was invented by Torricelli, a celebrated philosopher in Italy, the intimate friend of Galileo; but, unfortunately, the latter died three months after the former became his friend and associate. Torricelli himself died at the early age of forty, and thus the great expectations he had raised were crushed; yet the experiments he began were not neglected, and have been considerably improved on since his time. Like all first attempts, little accuracy was produced by the barometer invented by Torricelli: he formed it with a pipe sixty feet long, which being immersed, and suspended, in a vessel of water, after the air had been extracted, the water rose thirty-four feet in the tube by the pressure of the external atmosphere on its surface. This instrument being very inconvenient, induced him to attempt another, in which he used quicksilver, a fluid so much heavier than water, that a smaller quantity answered his purpose.

To make a complete barometer, according to the most improved method, a tube of glass, about thirty-three inches long, should be filled with quicksilver, and then immersed in a bason of that fluid; when the mercury in the tube will fall to about thirty inches, leaving a vacuum in the top of about three inches; and according to the state of the air, such will be the rise and fall of the mercury, between twenty-eight and thirty-one inches. This instrument has been employed to ascertain the densities of the air at different heights from the earth; for the quicksilver rising by the weight of the atmosphere, where that is lighter the depression, and where heavier the elevation, of the mercury in the tube, will express the various degrees of density. By this experiment, philosophers have discovered the air to be denser in the lower than in the higher regions of the atmosphere; for the quicksilver rose higher in a valley than on an elevation: and by observing the variations this instrument exhibited in ascending a mountain from its base, they estimated what must be the probable height of the whole atmosphere. The accuracy of their mode of estimating the weight of the atmosphere may be seen in the following experiment.

Experiment.

The barometer H, fig. 1, pl. 7, being placed on the little plate which communicates with the canal of the pump, and a receiver put over it, on exhausting the air from under the receiver, the quicksilver in the tube, having lost its support, will fall.

It may not be unpleasing nor useless to contemplate the mode by which mathematicians have made their calculations respecting the height of our atmosphere. Discovering, by the Torricellian experiment, that the whole weight of the atmosphere supported a column of water thirty-four feet high, a quantity weighing nearly fifteen pounds; also that, quicksilver being about fourteen times heavier than water, a tube one-fourteenth part the height of the tube of water being filled with it, the mercury was supported by the air; they weighed equal columns of common air and quicksilver, and found that quicksilver was 10,800 times heavier than common air: by which they were able to calculate very nearly the probable height of the atmosphere, allowing for its gradual decrease of density as it was further from the earth. I state this method as a matter of curiosity, but by no means wish to convey the idea of its being a perfectly accurate mode of estimating the whole height of the atmosphere. By the variation of the refractive power of the atmosphere, philosophers have ascertained its density at different heights with tolerable accuracy; and, according to their estimation, the rarity of the atmosphere is in geometrical, when the heights are in arithmetical, proportion: as thus, at the distance of seven miles from the earth it is four times rarer than at the surface; and at fourteen miles, sixteen times rarer; and so on.

The learned Dr. Horsley, Bishop of St. Asaph, treats this subject in a very beautiful manner, and with that strong sense and fine conception which characterise all his writings.—" If the atmosphere " of the earth reaches to infinite heights with finite density, those of

" Jupiter and every other planet will also reach to infinite heights
" above the surface of the planet with finite density. The atmosphere
" of every planet will therefore reach to the surface of every other
" planet, and to the surface of the sun ; and the atmosphere of the
" sun to the surface of them all.   All these atmospheres will mingle,
" and form a common atmosphere of the whole system. This common
" atmosphere of the system will be infinitely diffused, since the par-
" ticular atmospheres that compose it are so ; it will reach therefore
" to every fixed star: and, for the same reason, that of every fixed star
" will reach the central body of our system, and of every other system.
" The atmospheres of all the systems will mix; the universe will have
" one common atmosphere, a subtile, elastic fluid, which pervades
" infinite space ; and being condensed near the surface of every larger
" mass of matter by the gravitation towards that mass, form its pe-
" culiar atmosphere."   This reasoning is so beautiful, and speaks so
clearly the language of intelligent Wisdom, by according with the
harmony observable throughout the operations of nature, that I
feel perfectly satisfied with the opinion that it is so ; and I admire
the idea, which is certainly strongly expressive of the genius and
energy of the mind that conceived it.

Experiment on the Elasticity of the Air.

Of the amazing elasticity of the atmosphere, we have a familiar
evidence in the following experiment with a bladder, in which is a
small quantity of air.   I place the bladder in a box, A, fig. 4, pl. 8,
laying over it the weights B C D, each of which weighs two pounds ;
when, putting the whole under a receiver on the plate of the air-
pump, and exhausting the air from the receiver, the small portion
of included air in the bladder expands, and raises and supports the
weights, as seen at A, fig. 5, pl. 8; on re-admitting the air, the
bladder sinks to its former dimensions.

Experiment.

The glass A, fig. 6, pl. 8, has a bladder tied over it. I shall place this under a receiver, on the plate of the pump, and exhaust the air from it. On removing the external air from the upper surface of the bladder, the air included within the glass exerts its elasticity on the inner surface of the bladder, and bursts it in pieces; but with very little noise in comparison with what you heard when the external air burst the bladder, because in that instance the blow was given by a larger quantity of air, and in this experiment the sound is diminished also by being in a vacuum.

We have seen a fountain play by the pressure of the external air; we may now observe a similar effect produced by its elasticity.

Experiment.

The bottle A, fig. 7, pl. 8, is about two-thirds filled with water; the upper part of it is occupied by air: on placing it under a receiver, and exhausting the external air from the bottle, the expansive power of the air within it, acting on the surface of the water, forces it through the glass tube and jet B B in the manner of a fountain.

The hygrometer is an instrument used to ascertain the quantity of moisture in the atmosphere. Various means are employed for this purpose; but a small weight tied to a piece of catgut, and suspended from a prop, hanging perpendicularly to the horizon, with a scale of degrees placed behind it, will answer tolerably well. In damp weather the catgut shortens, and in dry it lengthens; which changes are effected by the twisting and untwisting of the cord, its bulk being increased by the presence, and diminished by the absence, of the moist particles of the atmosphere.

All bodies have air in their interstices, and it may be extracted
from them in four different ways :—by heating them ; by cooling
them ; by dissolving them ; and by placing them under the receiver
of an air-pump. By heating a substance, we expel the air contained
in its interstices or pores, the heat dilating the air so much that the
same space will not contain it ; for unless the materials expand, and
in the same proportion with the air in their pores, that fluid must
necessarily escape from them. If a substance, by the condensation
of the air, be considerably cooled, its parts becoming compressed
more closely together, the air is squeezed from it as water from a
sponge. Air may be disengaged from a substance by solution;
for this operation separates the parts of a body from each other,
consequently the air must escape, as we see it from sugar when
dissolving in water, the water rendering the effect visible. The air
may also be disengaged from a substance by placing it under the
receiver of an air-pump; for by diminishing the weight of air by
rarefaction, the air within the substance is dislodged, while the external pressure is removing.

Experiment.

However close the texture of wood, it has air in its interstices,
as we shall perceive from the following experiment. Placing a
piece of dry wood, A, fig. 1, pl. 9, within an open receiver in a
vessel of water, and laying my hand on the top of it to cut off its
communication with the external air; on lessening the density of
the air in the water, that which occupied the pores of the wood
passes into the water; and on withdrawing my hand, the external
air will rush through the wood with great velocity. The vessels,
or canals through which the fluids pass in wood while it is growing, and in which the particles of air reside when vegetation has

ceased, cause joiners' work to vary in its dimensions according to the state of the atmosphere, and occasion the crackling noise heard in wainscoting and other thin portions of wood when the state of the air changes.

As the regulation of the temperature of the air in apartments is of consequence to us, and many inconveniences arise from ignorance of the most familiar effects of the atmosphere, I trust I shall be excused for introducing some remarks on the construction of chimneys, and the effects to be either produced or avoided by our knowledge of the properties and tendencies of heat. The healthiest people are most liable to injury from heated apartments, where there is not a sufficient current of fresh air to supply the expulsion of that fluid by heat: habit renders them so; for as health is best maintained by exercise in the open air, the cool quality and strong impressions of it so respired by the robust become necessary to them; and when either is diminished, such persons particularly feel the bad effects of the deprivation. Fire is supposed to purify the air of a room, and I conceive by inducing a strong current, which forces the vitiated air up the chimney, and thus refreshes the apartment.

OF THE CIRCULATION OF AIR IN CHIMNEYS.

IN general, chimneys draw well that are contracted near the body of the fire, and that are carried up high and winding; for the air nearest the fire must be the hottest, and when that body of air is most heated, it increases the force of the current of air. Towards the top of a chimney a winding form is best to prevent the effects of sudden gusts of wind, which act powerfully on the smoke of a whole chimney built in a straight column.

We all know that fire is increased by a blast of air, as we see the effect produced on it by a pair of bellows. Thus, the necessity there is that a room should admit a sufficient quantity of air, to keep up that quick circulation which is required to supply the consumption of the vital principle, is evident. Sometimes a chimney will draw very well on first lighting a fire in it, but will smoke afterwards. This effect is produced by the rarity of the air in the apartment, which is become of nearly the same temperature as that in the chimney; and, therefore, will no longer force up the smoke. When a fire smokes on first lighting, but not afterwards, this arises from the magnitude of the cold column of air in the chimney, which does not become warm till the fire has acted on it a considerable time. We should never suffer a draught of air in any other direction than that of the chimney; for if we do, the draught will draw the air from that part into its current, and cause the smoke to fall into the room. Smoke is a vapour which is supported by air; therefore it will fall when it loses that support. The experiment of placing a burning taper under a receiver shews this; for when the air is too light to support the smoke, it falls on the plate of the pump; whereas at first it rises, and is supported by the air. Chimneys in stacks generally draw best; as they are not so much cooled by the external air. A building higher than a chimney causes eddy winds, which prevent the rise and dispersion of the smoke.

There is always a draught of air in a chimney, even when there is no fire in it: changes in this current of the air have been observed, and may be accounted for on rational principles. In the summer a regular draught passes up a chimney in the night, and down it in the day time, as there is at those times great difference in the state of the air: but a chimney inclosed within the walls of a house, is not sensible to either the influence of the sun's beams or the deprivation of them, but is in a medium state between the two extremes; so that at night, when the air of an apartment is colder than that within the chimney, it forces the warmer air upwards; and in the day time, when the air of the room is hotter than that in the chimney, the colder air, by its density, descends into the room. These effects may be perceived by placing a lighted candle in a chimney both in the night and in the day time.

We have seen that the air has weight, that it may be rendered lighter, and that its dimensions may be much increased. It may also be rendered specifically heavier than in its natural state, by condensation.

Experiment on the Condensation of Air.

The metal vessel represented by fig. 1, pl. 10, being partly filled with water, and the remainder occupied by air in its natural state, we will endeavour to introduce a greater quantity of air into the space now occupied by that fluid; but, previously, we will contemplate the machine employed to effect this purpose.

Experiment.

Fig. 4, pl. 9, is called a condensing syringe. It has a solid moveable piston, A. On lifting up the piston-rod, a vacuum is left below, till it arrives at a small hole near the top of the barrel at B, through which the external air presses to restore the equilibrium. There is a valve at the bottom of the syringe, opening only downward; therefore, when the solid piston is depressed, the air that is admitted into the barrel through the perforation, B, is forced into any vessel to which the syringe is screwed; and, by continuing the operation, we may condense the air beyond the power of almost any vessel to contain it. But in the following experiment it is necessary only

to condense it sufficiently to make the water ascend in that proportion and to that height which will be found pleasing and convenient.

On screwing the syringe to the top of the stop-cock at A, fig. 8,
pl. 10, which is open for the purpose of forming the communication,
after pumping till the air is sufficiently condensed, we shut the cock,
and unscrew the syringe, and put on a jet in its place. Then, on
opening the communication, a beautiful fountain is produced by the
action of the condensed air on the surface of the water within the
vessel.

Various jets are contrived to give variety to the fall of water from
this fountain; see figs. 1, 2, 3, 4, pl. 10. The one now supposed on
the top of the machine supports a cork ball, which rises and falls
with the water, and revolves on its centre. The beautiful effect of
the rainbow may also be exhibited, by letting the sun shine through
the water thrown from the jet, fig. 6, which clearly explains the
cause of that striking phenomenon. B, fig. 2, pl. 10, is an arm to
screw on occasionally, so constructed as to give variety of direction
to the water.

The experiment I have just exhibited is a sufficient demonstration of the power of condensed air; yet, to convince you that the
increased power imparted is owing to a greater quantity of air being
introduced into a given space, we will weigh a flask full of air in
its natural state; and, screwing on the syringe, condense the air
by introducing a greater quantity into the same space; when, on
again weighing the flask, we shall perceive the degree of condensation by the difference of weight produced by that means. The
effect of condensed air, as shewn in the fountain, serves to convey a
pretty general idea of its amazing force when greatly increased.

It is by the pressure of the atmosphere that we raise water from a well by means of the common pump; which, like many other familiar operations, is a philosophical experiment. We know that the pressure of the atmosphere is capable of raising and supporting thirty inches of mercury, which are equal to the weight of a column of water thirty-four feet high, and no more. Therefore, water cannot be raised by this pump to a greater height than thirty-four feet; nor is it constructed to raise it so high, because the weight of the atmosphere is sometimes much lessened. The wooden trough, A, fig. 2, pl. 9, represents a well from which water is to be raised, and the cylinder A, fig. 2, the body of the pump: B is the moveable piston, in which, at c, is a hole covered by a piece of brass, called the valve, which opens upwards only: there is also a valve at the bottom of the barrel, at D, opening in the same direction.

Experiment.

When I raise the piston, I bring up a column of air out at the top of the pump. The air in the barrel is thus rarefied, by which means the pressure of the external air acts more powerfully on the surface of the water in the well than that from the top of the pipe. By degrees, a perfect vacuum is produced within the barrel; and the water consequently rises to the top, and then runs out at the spout. This operation would require no labour were it not for the friction of the piston, and the pressure of the external air on the surface of the water above it, which renders it heavy.

Experiment.

E, fig. 2, pl. 9, is called a forcing-pump; it is used for raising water to a considerable height, by the condensation of air. There is no valve to the piston, A, of this pump, which, being solid, when drawn up reduces the air in the barrel, by which means the water rises above

L

the valve c, and enters into the bottom of the barrel B, whence it passes through the tube D into the air-vessel E, by the valve F, which opens upwards only. The operation thus far is the same as is performed by the common pump. When the water in the air-vessel rises above the bottom of the pipe G, as the air is prevented from escaping when the bottom of the pipe is covered with water, the air in the upper part of the barrel becomes condensed by the water pressing it into a very small space; which density causes it to exert an equal pressure on the water round the pipe, so as to project it through the jet H to a considerable height. The smaller the space into which the air is compressed, the higher the water will be thrown from the jet. This machine is applied to many useful purposes. Some have been contrived that have raised 140,000 hogsheads of water in a day.

The following very easy experiment will convince those who have not an opportunity of trying any other, of the power of condensed air.

Experiment.

After filling a bladder with your breath, tie a syringe to it, and on condensing the air the bladder will burst in pieces, by the superior force the air will exert from an increase of its density or quantity of matter.

Air-guns are discharged by the condensation of the air. Fig. 4, pl. 10, represents a section of one, which shows the interior construction of this machine, and will serve to give us an accurate idea of its nature and mode of action. The air-gun is made of brass: it has two barrels; the inside one, A, of a small bore, from which the bullets are discharged; and B B, a larger barrel on the outside, encompassing the smaller one. In the stock of the gun is a syringe, c d, having a rod, l, that is drawn out to take in air, when on pushing it in again, the piston e drives the air before it through the valve f, into the cavity between the two barrels. The ball g is put down into its place by a ram-rod, as in other guns. There is a valve at h, which being opened by the trigger k, the condensed air rushes on the ball, and discharges it with great force.

How does each new subject raise our admiration of the kind, provident and protecting goodness of our great Creator! Surely no one can be so blind as not to perceive in the wonderful processes of nature a regular arrangement of causes and effects, produced by infinite Wisdom and Beneficence. How greatly, then, ought we to rejoice in every opportunity that enables us to contemplate our Creator in his works! This exercise of our reasoning powers strengthens our judgment, and elevates our ideas of religion and morality, placing them in their proper rank—the first in our esteem and admiration. Through the properties of air we have already investigated, we trace the hand of an all-wise Providence liberally bestowing benefits on creatures dependent on his goodness. Yet the unthinking many disregard these evidences, and, till roused to reflection, feel not the gratitude for them which must glow in the breast of the natural philosopher.

The desire of information is generally the first incitement to philosophical researches. Sometimes particular circumstances and conditions of life awaken our energies and stimulate our researches after philosophical truth, to console us under temporary disappointments, or to assist our temporal concerns. Whatever be the incitement, the result is certain improvement in all that is amiable in disposition and praise-worthy in practice—

——————————————— for the attentive mind,
By this harmonious action on her powers,
Becomes herself harmonious : wont so oft
In outward things to meditate the charms
Of sacred order, soon she seeks at home
To find a kindred order to exert
Within herself this elegance of love,
This fair inspired delight; her tempered powers
Refine at length, and every action wears
A chaster, milder, more attractive mien.   AKENSIDE.

PLATE XI.

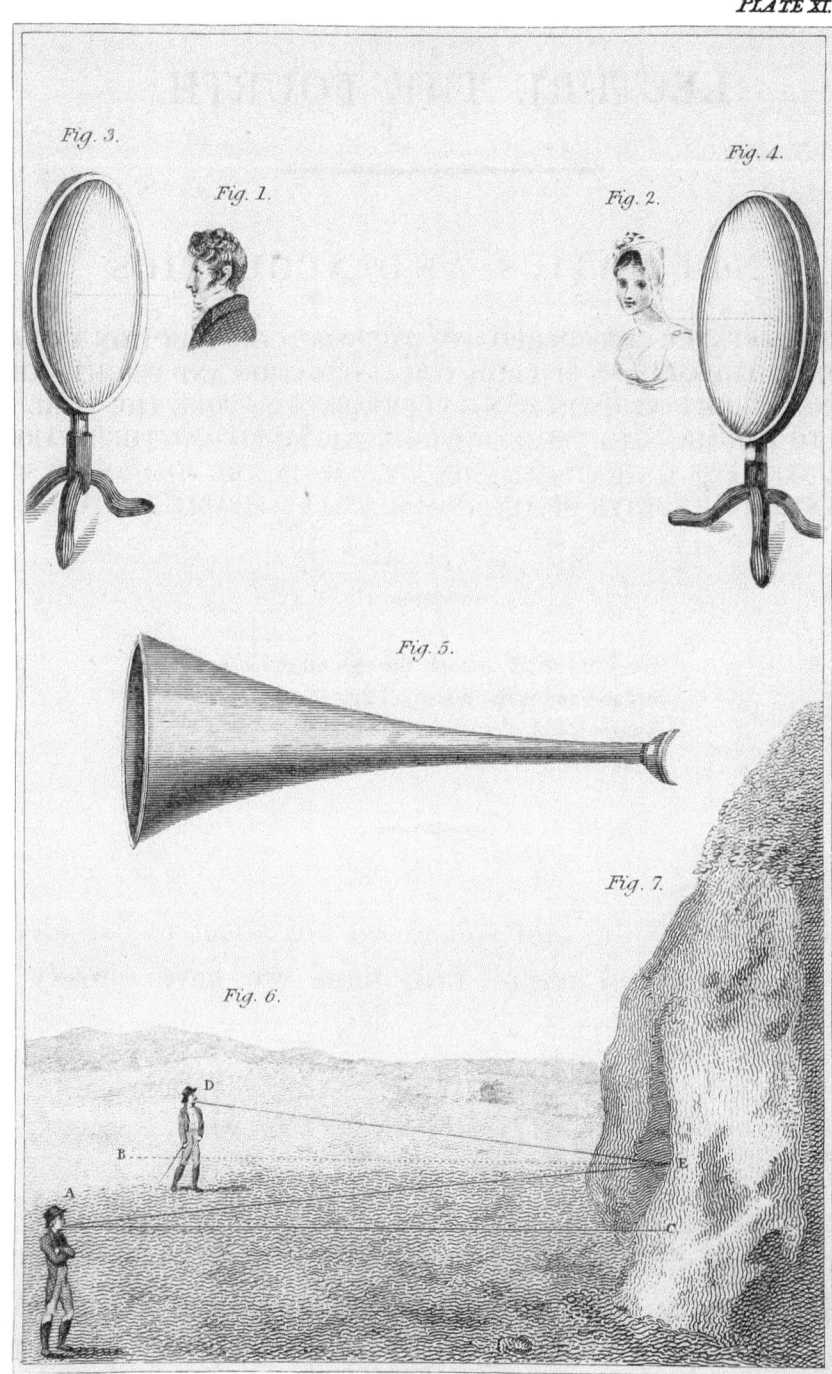

Fig. 3.

Fig. 1.

Fig. 2.

Fig. 4.

Fig. 5.

Fig. 7.

Fig. 6.

M.Bryan design.   T.Noble delin.   H.Mutlow sculp.

LECTURE THE FOURTH.

PNEUMATICS AND ACOUSTICS.

THE AIR'S RESISTANCE CONSIDERED.—OF THE WISE CONSTRUCTION AND ADAP-
TATION OF THE ORGANS OF BIRDS FOR RESPIRATION AND FLIGHT.—OF THE
ORGANS OF RESPIRATION IN MAN.—VEGETABLES RESTORE THE VITAL PRIN-
CIPLE TO THE AIR.—THE CAUSE OF WINDS EXPLAINED, AND THEIR VARIETIES
ACCOUNTED FOR ON NATURAL PRINCIPLES.—AIR THE MEDIUM OF SOUND,
AND A CHIEF SUPPORTER OF ALL ANIMAL AND VEGETABLE EXISTENCE.

The Lord of all, himself through all diffus'd,
Sustains and is the life of all that lives.
Nature is but a name for an effect,
Whose cause is God.   COOPER.

THE properties of the atmosphere we are about to consider, are
equally powerful and useful with those we have already contemplated.

ON THE RESISTANCE OF THE AIR.

The resistance of the air, is a property of it much attended to by
man, in many of the operative springs of action, to which it becomes
subservient in a greater or a less degree, according to the force of
the machine employed, and the manner of using it. The air's resistance is in proportion to the surface opposed; for the same weight

may be more or less resisted according as it presents a greater or a less surface to this opposing medium.

All bodies moving in a fluid are retarded by it, because of the resistance of its parts, which diminishes the force of the moving body : and the larger the surface opposed, the greater its retardation; as the opposition of more parts of the fluid must be overcome, and consequently the momentum of the body in motion will be lessened in that proportion. We may be convinced that the air, in its independent state, is a resisting body, by the following popular experiment.

Experiment.

We will take the tall receiver, fig. 3, pl. 9, and place it on the plate of an air-pump. A fits the top of the receiver, on the appendages of which, B C, a guinea and feather are to be laid at the beginning of the experiment. Letting them fall in air, the guinea will arrive at the bottom of the receiver before the feather; but on extracting the air from under the receiver, we shall see them both reach the bottom of it at the same instant.

The resistance of one fluid may be compared with that of another; therefore, when mankind discovered this property of the air, they applied themselves to ascertain the effect of resistances in other fluids; on which known principle they constructed machines the best calculated to overcome them, either by their form or mode of action. Hence, the oars a waterman employs to drive forward his vessel, are made broad and flat in one of their dimensions, and sharp in the other. He causes them to act on the broad side when he wishes to force his boat on, and lifts them out of the water edgeways that he may have less resistance to overcome.

ON PROJECTILES.

LET us consider for a moment the great advantages derived from the knowledge of the resistance of air, in improving the theory of projectiles. The theory of projectiles applied to military purposes is the calculation of the resistance of the atmosphere, in order to know the course a ball will take through the air, and the distance it will range. We might suppose that heavy balls, passing through the air with great velocity, would not meet with any sensible resistance: but this is not the case in fact, as the greater the velocity the greater resistance it creates ; for it is known that a ball fired from a cannon which will pass through two miles in air, would range forty miles in vacuum. It is unnecessary for us to enter into the consideration of projectiles; I merely mention them as being connected with the subject of resistances. I have treated largely on the laws of motion and gravity in my lectures on Astronomy, so far as relates to the revolutions of the bodies which compose the solar system: but that is foreign to my present purpose; therefore I shall only state the circumstances that are considered in regard to projectiles when employed on terrestrial subjects, namely, the impressed force, the resistance of the air, and gravity. These are all allowed for in the ballistic art, which renders the theory of projectiles more difficult than any other mathematical calculation; for, though the effects of gravity are always alike at the same place and at the same distances from the earth, yet the density of the air varies considerably at different times. Of the difficulty attending this calculation we may form some idea, when we know that in gunnery a piece of ordnance must have a certain elevation to enable it to convey a ball to a particular spot; that the increasing effects of gravity on it, as it approaches the earth, must be ascertained; the time of its performance must be known; the quantity of the impressed force, and the air's resistance at that time, must be estimated; in order to make the ball reach. a given spot with a determined direction and velocity.

The resistance of the air is employed by our Creator for the benefit of all his creatures, and is peculiarly useful to the feathered race. By striking the air with their wings, which are supplied with muscles adapted for that purpose, the resistance of that medium pushes them forward, their tails serving them as a rudder to regulate their course. Such birds as have short tails instinctively stretch out their legs behind them, to assist their steerage and preserve their balance. The breast-bones of birds are sharp, and gradually diminish from their spine; which enables them to divide the fluid medium, and pass through it with great rapidity. Birds of quick flight, such as swallows, have small heads terminating in sharp-pointed beaks, small bodies, and large wings; but birds of slower and less distant flight, have larger bodies, and fewer feathers. Sparrows and linnets fly only by starts, supporting themselves but for a short time in the same position. Their wings opposing only a small surface, the reaction is not sufficient to sustain them long, without frequent renewal of the velocity or exertion.

The reaction of the air alone is not sufficient to account for the direction, &c. of the flight of birds: it is by the curious mechanism of their wings that they are able to support themselves, and to vary their flight, as we may readily conceive; for if it were performed and effected by strokes in one plane only, what was gained one moment would be lost the next. But the wise Contriver of all things has supplied the plumy race with a curious and wonderful machinery to effect these purposes. The external part of the wing is convex, and the feathers are so disposed, the muscles and joints of the pinions so arranged and allotted, as to enable birds to shift their position by a semi-rotary movement, and also to strike the air with a broad surface, in order to take all possible advantage of its resistance; and to raise the wing edgeways, that they may have the less opposition to overcome, and to prevent that action of the air on them in rising which would impede their flight : it was probably from observing this circumstance, that the waterman learnt what is called to feather his oar.

We find the investigation of the subject of air important as we trace the uses of that fluid in the common operations of nature and of art ; but it becomes more so when considered in a religious view, as an evidence of Omniscience, of which the closer it is examined the brighter the signatures will appear, as you will confess when I show you the adaptation of the organs of different species of animals to the purposes of respiring this fluid medium.

Anatomists have discovered that birds respire by means of lungs so constructed, that the air passes through them to every part of their bodies. This diffusion of the air-vessels in their bodies prevents the inconvenient effects, that would otherwise arise from the great action of the air on them, in passing quickly through it ; for were the air confined in birds as in man, the celerity of their motion would produce suffocation.

The resistance of the air increasing in proportion to the quickness of a body's motion through it, and also to the largeness of the surface opposed ; if a man could move in the atmosphere with the same velocity as a bird, not being provided with similar organs of respiration, and presenting a surface so much larger to the resisting

M

medium, he would soon be destroyed by the powerful impressions of the air.

Some fishes respire like man, which obliges them to come to the surface of the water at intervals, to throw out the foul air, and take in a fresh supply. Other fishes respire through gills, both water and air. Most insects have organs of respiration, though not lungs, like men, quadrupeds, &c.

It would be impossible to describe the various contrivances for respiration in the different orders of beings; therefore, I shall endeavour to exhibit as clearly as I can, from the authority of the excellent Dr. Paley, the organs of respiration belonging to the highest order of animal nature only; inferring from thence, and from what I have related, that all creatures are endowed with the pulmonary powers and capacities particularly adapted to each species.

The windpipe, through which we respire, communicating with the mouth, is furnished with a valve; over this valve, which opens and shuts twenty times in a minute, the food passes, without entering the windpipe; for when the food approaches, the valve instinctively closes. The lungs are two viscera, contained in the thorax, or chest; and are there fenced with bones and sinews. The ribs, being turned into a regular arch, are moveable on a kind of centre, which assists the action of respiration, and forms a secure lodgment for the lungs and heart, the most important organs of animal life. The heart is placed between two soft lobes of the lungs, and sustained in its place by muscles, cartilages, &c.

The lungs are considered as right and left; they have a broad base, and terminate in a point at top. The right lung is divided into three, and the left into two, lobes.

The substance of the lungs is composed of numerous small bags of air, divided from each other by little cellular cavities, through which blood-vessels are distributed. These cells are variously figured, and of different dimensions. The pulmonary artery and veins are distributed throughout the whole lungs; the latter encompassing each bag like a net. The windpipe descends into the thorax, and unites with the pulmonary artery and the veins; altogether composing the spongy substance of the lungs. The bones, cartilages, and muscles, which form the cavity of the stomach, are so arranged that the chest can be contracted and enlarged at pleasure, for the purposes of inhaling and exhaling the vital air; which operations are performed partly by the reciprocal elevation and depression of the ribs, and partly by raising and depressing a muscular partition which divides the thorax from the abdomen. This partition rises convexly upwards, in the stomach or chest; but it sometimes becomes a plane surface, for the purposes of respiration. On drawing air into the lungs, the chest expands; and in throwing it out, the chest contracts.

Such a variety of organs being necessary to respiration, and the mechanism of them being very complex, no wonder that, after all the attempts which have been made, no instrument has yet been invented to exhibit with tolerable accuracy this faculty of animal nature.

The machine called a lungs-glass is a very imperfect representation of the operations of breathing; and I cannot with satisfaction exhibit it as such, because its deficiencies lessen the grandeur of the contrivance employed by our Divine Creator to effect his purposes. To confirm this truth, let us contemplate only two of the circumstances concerned in respiration, which cannot be exhibited by this machine—the blood from the heart poured into the lungs to assist the faculty of breathing, and the pulsation of the heart.

How can a glass bottle, in which only a single bladder is suspended, even assisted by the action of the air-pump, exhibit these things? Impossible! Let our thoughts then dwell on the wonderful combination of causes that produces the indispensable action of the lungs, which curious mechanism and economy excite equally our gratitude and admiration.

The vital principle depends on the state of the air we breathe; for if we inhale very impure air, so as to obstruct the action of the lungs, that of the heart ceases; or if we cannot discharge the particles of vitiated air from the lungs, the same effect takes place, and we expire. When our lungs are in a perfectly healthful state, by inspiration we inhale the wholesome atmospheric air, and by expiration discharge air which has been contaminated. This fact has been fully established by too many cruel experiments; one only should be sufficient to satisfy a reasonable being. I highly disapprove of that exercise of our power over the inferior orders of animated nature which inflicts pain, unless essentially necessary for their benefit or ours. Nor can I think that man truly humane, who does not feel for suffering nature, however insignificant the object that endures torture, or feels the pangs of death. Melmoth, in his note on the humanity of Pliny's temper, justly, humanely and beautifully observes—" that true benevolence extends itself through the whole compass of existence, and sympathizes with the distress of every creature. Little or depraved minds may consider compassion to the inferior order of beings a weakness, but it is the true genuine demonstration of a noble nature." Homer, whose appropriation of character is always correct, did not consider it unbecoming the dignity of a hero to shed tears at such distress; for he has drawn a most affecting picture of Ulysses weeping over the expiring moments of his faithful dog.—

> Soft pity touch'd the mighty master's soul;
> Adown his cheek the tear unbidden stole
> Still unperceiv'd; he turn'd his head, and dried
> The drop humane. POPE'S ODYS. 17.

But, to return to the subject which produced this digression.— As animals, by breathing, contaminate the air, it is necessary to keep apartments properly ventilated, in order to expel the imparted noxious quality, and to procure a fresh supply of the vital principle. Air is as necessary to vegetables as to man; for if a plant be deprived of air, though it have all the benefit it can derive from the sun, it will die.

The great quantity of noxious or impure air absorbed by plants in their natural state, is by the Providence of God emitted from them pure. What a sublime and beautiful idea does this impart! The Almighty rendering one part of the creation, without injury to itself, beneficial to the other; for, by this absorption and regeneration, the atmosphere is rendered fitter for the respiration of man.

Dr. Priestley communicated this knowledge, after making various experiments to authenticate it; one of which will be sufficient to convince us of the reasonableness of his inferences. Putting sprigs of mint in air, that had been previously deprived of its vital principle, and in which a candle would not burn, he perceived the mint throve exceedingly, and that afterwards the air supported the flame of a candle; from which he inferred that the mint had restored the vital principle to the air: this confirmed his opinion, that plants, instead of injuring the air by vegetation, as animals do by respiration, restore the vital principle to it, rendering it fitter for animal existence: this knowledge he communicated to Dr. Franklin; on which the doctor sent him the following letter.—" That the vegetable " creation should restore the air which is spoiled by the animal part " of it, looks like a rational system, and seems to be of a piece with " the rest. Thus, fire purifies water all the world over; it purifies it " by distillation, when it raises it in vapor, and lets it fall in rain; " and farther still by filtration, when, keeping it fluid, it suffers that " rain to percolate the earth. We knew before that putrid animal " substances were converted into sweet vegetables, when mixed with " the earth, and applied as manure; and now it seems that the same " putrid substances, mixed with air, have a similar effect. The " strong thriving state of your mint in putrid air, seemed to show " that the air is mended by taking something from it, and not by " adding to it. I hope this will give some check to the rage of " destroying trees that grow near houses, which has accompanied " our late improvements in gardening, from an opinion of their " being unwholesome. I am certain that there is nothing unhealthy " in the air of woods; for we Americans have every-where our " country habitations in the midst of woods, and no people on earth " enjoy better health." The above authorities are sufficient to confirm our opinion on the subject. Those who wish to be acquainted with all the properties of the various kinds of air which may be formed by combinations, and the elementary principles of the common air of our atmosphere, I beg to refer to Dr. Priestley's ingenious experiments, published by Johnson, St. Paul's Church-yard; contenting myself with mentioning those qualities of the atmosphere which may be termed mechanical.

The air we breathe is composed of a mixture of mineral, animal, and vegetable exhalations, arising from the earth and the bodies on its surface. It is formed by the volatile parts of these bodies, raised by the heat of the sun, or driven from them by fermentation, &c. The atmosphere is of an elastic, weighty, and penetrating nature; susceptible also of expansion; which latter property demonstrates the repellancy of its parts.

The attraction of the earth, and the increased incumbent pressure, make the aerial fluid more dense near its surface. The atmosphere absorbs moisture, supporting great quantities of water: but it will imbibe a certain portion and no more; for when perfectly saturated, the watery particles descend in rain, hail, &c.

Vital air, which supports animal and vegetable existence, composes one-third part of our atmosphere. Dr. Priestley asserts that this principle is given out by plants, during the influence of the solar light; but in darkness is retained in them.

We may suppose the influence of the sun necessary to produce that dilatation of the parts of bodies which is essential to the emission of the vital fluid; and therefore in darkness we may consider it to be in an inactive state. This principle in the air is not distinguishable from the other qualities, but by its effects; it naturally exists in the atmosphere in a sufficient degree for our free respiration, and may be produced artificially, or at least a purer quality may be imparted to a limited quantity of air, which is made evident by its supporting combustion for a longer time than the same quantity of common air; and the difference will be in proportion to its greater degree of purity. This air is produced from nitre, acids, metals, and all earths; which proves it is derived from the earth, and we may suppose it is communicated to the atmosphere through the pores of vegetables, &c.

The state of the air has considerable effect on our bodies: in fine weather, when the clouds rise high, or the atmosphere appears perfectly clear, we feel light and cheerful, because the density of the air at that time acts powerfully on our bodies, and induces a proportionate elastic force in the air within them: but in wet weather, and when the clouds are low, the air is light and its impressions feeble; consequently the reaction must be the same, and therefore we feel dull and indolent.

We are convinced, both by reason and experiment, that though the air is invisible, it is a powerful agent in nature, and a principal medium of existence. The atmosphere is heavy and elastic; it supports water and fire, is expanded by heat and condensed by cold, and always endeavours to place itself in equilibrium.

These known properties readily explain the cause of winds; for when any portion of the atmosphere is rarefied by heat, the colder air rushes in to restore the equilibrium. Within the torrid zone, there is a constant current of fresh air from the east, which is produced by the earth passing the sun from the west to that point. That spot of the earth immediately under the sun is more heated than any other part, and consequently communicates a greater degree of warmth to the atmosphere in immediate contact with it; this causes a rarefaction, which leaves room for a current of air from the east to rush in and restore the equilibrium. The north wind combines with it, when the sun is on the north side of the equator; and with the south wind, when it is on the south side of it. This direction of the current of air is only perceivable at a distance from the land; for near coasts various causes produce irregularities in the winds. A breeze always sets in from the sea in the day-time; the sun heating the land more than the sea, the atmosphere over it is more heated, and consequently the current of air is from the sea while the sun is above the horizon. But at night the current is from the land; for after sun-set the air on the earth sinks, and the superabundant quantity returns to the ocean.

The sandy deserts of Barbary impart a great quantity of heat to the atmosphere over the Mediterranean, and the neighbouring lands. This hot air, being loaded with vapour, is rendered quite suffocating; and the wind which passes over those deserts to the coasts of Greece and Turkey, called the sirocco, is sometimes intolerable to the inhabitants of those places.

In the Indian ocean there are winds called monsoons, which blow from the south-west. Their effects are most evident when the sun is in northern declination, and nearly vertical to them; which is from April to September. From September to April, when the sun is over this ocean, and the islands of Borneo, Java, and New Holland, the current is from the north-west, because the atmosphere of these islands is more heated than that over the ocean. The currents of air or winds in different places are various. These variations are produced either by exhalation from the earth, by volcanoes, difference of soils, or a greater or less quantity of the sun's rays; all of which, under certain circumstances, produce hurricanes and tempests. This agitation of the atmosphere is however necessary, to preserve its purity; for stagnated air soon become unfit for respiration.

When the effluvia from the earth become strongly impregnated with alkaline and acid particles, a great fermentation arises, and flame is produced, which occasions the meteors sometimes seen in the air. This effervescence is evident on mixing acid and nitre together, which immediately ferment. Sometimes the inflammable matter will flash in the air, without doing any harm; but when powerful, it explodes with great violence. Such combustible materials effervescing in the earth, produce earthquakes. Volcanoes are also occasioned by a great internal commotion arising from the fermentation of various elementary particles; but principally from water suddenly rushing into the apertures, and being as suddenly converted into steam.

OF SOUND.

WE may now proceed to consider that property of our atmosphere, which affords us the delight arising from musical harmony, and the soothing, the animating, the captivating voice of friendship and love, in all their dearest ties of affinity and consanguinity.

The densest air is the best conductor of sounds; which is evident on placing a bell, fig. 2, pl. 10, within a receiver: on exhausting the air, as we continue to reduce its quantity, the sound becomes fainter and fainter. We may conceive this to arise from the pulses of air reacting with less force, in consequence of its becoming less dense.

The effect of sound is greatest when produced by solid bodies: when we strike a piece of metal capable of vibration, which is composed of very dense particles, the sound is heard very clear on our auricular organs.

As air is the natural vehicle of sound, we will consider the manner in which it receives the impression and conducts it to the ear. The particles of air move with the utmost facility among themselves; when, therefore, the air receives an impulse, the particles in immediate contact with the issuing sound act against the neighbouring ones, and these against the next, and so on, producing a continued effect from the point whence the sound proceeds to the ear which receives it.

Musical sounds, proceeding from a stringed instrument, depend on the vibration of the strings; the more acute arise from quick, and the graver from slow, vibrations; the time of each vibration depending on the length, magnitude and tension of the strings; and thus producing the variety in musical tones. If one musical string be twice the length of another, the shorter performs two vibrations while the longer performs only one; which is according to the laws of motion. This vibratory motion communicated to the air is in the form of waves, one wave surrounding another on all sides.

The science of sounds is divided into two parts: Diacoustics, which treat of sounds directly striking the ear; and Catacoustics, which relate to reflected sounds : and both are sometimes expressed by the single term acoustics. The notes of stringed instruments are generated and subdivided into different gradations of sound, according to their lengths, breadths, and tensions ; and into loud or soft, sharp or flat, with all the intermediate degrees. Sound is communicated in time, as you may perceive on the discharge of a cannon, for you will see the fire before you hear the report.

Sound, by being condensed in a tube, is rendered audible at a great distance; therefore, by means of pipes, which confine the sound, the voice may be heard considerably beyond its natural limits. Hence have arisen various deceptions. The condensation of sound has limited effects with the condensation of light, in increasing the natural powers. When a person speaks in a trumpet, the large waves formed at the wide end of it are compressed at the axis by the reflecting surface inside the tube; and, passing to the ear in that state, a greater effect is produced than by the usual mode of conveyance. If two trumpets are fixed in situations opposite to each other, even at the distance of forty feet, the sound of the lowest whisper spoken at the mouth of one of them, will pass to the other, and may be distinctly heard. The similarity between the effects of condensed light and sound is evidently proved by experiment. If we place two concave mirrors, or surfaces, made of glass or any reflecting substance, at the opposite ends of a large apartment, and a person stand at the focus of one, as at A, fig. 1, pl. 11*, and another person at the focus of the other, B, fig. 2, pl. 11, they may converse in the lowest whisper, which will be to each perfectly audible.

Experiment.

Suppose A, fig. 1, pl. 11, to be a person speaking at the focus A of fig. 3. The voice, issuing in a spherical direction round him, and striking against the concave surface of the reflector fig. 3, is reflected, and impinges on the reflector fig. 4, being condensed at its focus B: and it makes a greater impression on the ear than if it were not aided by these auxiliaries, or condensers of sound.

The speaking-trumpet is of early invention; for we are told by historians, from the authority of a manuscript found in the Vatican

---

* The focus of a concave mirror is at one-fourth the diameter of a sphere of which the concavity of the mirror is an arc.

at Rome, that Alexander had a horn through which he gave orders to his whole army ; that its greatest diameter was nine feet and a half; and that it was heard at the distance of twenty-five miles. This, however, is supposed to be an exaggerated account ; yet we may believe that some such instrument was used by Alexander, and for the purpose mentioned. The speaking-trumpet is a long tube, gradually increasing in width from the mouth-piece to the other extremity, inclining constantly toward the outside. The small end of this tube is adapted to the form of the mouth, and has two lateral projections, one on each side, which press on the cheeks, in order to retain the sound. Fig. 5, pl. 11, represents this instrument.

The effect given to sound by the speaking-trumpet is produced by condensation, thereby rendering its impulses stronger, and thus also conveying it to a greater distance. When a person speaks at the mouth of the trumpet, all the motion that would be communicated to a spherical mass of air, suppose of forty feet radius, is confined within a very small space, namely, the tube of the trumpet; now, if the trumpet be only a hundredth part of the whole sphere the sound would fill, by its concentrated effect it is augmented about a hundred times, and consequently will be heard at a hundred times the distance that it would be without the aid of this instrument.

To produce sound, the body struck must be susceptible of vibration. Such substances as are of a hard and brittle nature are the most sonorous. In wind instruments, the vibrations depend on the size of the aperture whence the sound issues, and the strength and velocity of the percussion.

Though air is a necessary medium of sound, yet a small quantity only is necessary to convey it to the ear; and it is observable that sounds are heard best close to the vibrating body which emits them. The pulsations of sound are transmitted in circles, forming spheres round the point from which they issue. The softest sound is heard in as short time as the loudest; but the more distant we are from the generating point, where the imparted motion must be the greatest, the fainter the sound becomes. It has been ascertained that sound passes through a space of about 1142 feet in one second of time. From which knowledge it is easy to calculate the distance of a thunder cloud, by observing a flash of lightning, and noticing the time that elapses between the lightning and the sound of the thunder.

Sound is reflected in the same direction as light from a mirror, that is, the angle of reflection is equal to the angle of incidence.

Experiment.

Suppose A, fig. 6, to be a person speaking. The sound, issuing in circles round his head, strikes against any object within its sphere of action. Let fig. 7, pl. 11, be part of a mountain within that sphere; the line A C representing both the incident and reflected sound; for the sound is reflected in the same line of direction from C to A. Supposing the incident sound to proceed from a person at D, striking the rock at E, it will be reflected from E to A; for the angle of reflection, A B, will be the same as the angle of incidence, B D. These effects are called echoes; but there must be a certain time and a determinate distance, and no object intervening, to render an echo distinct. In my opinion the best echoes proceed from various reflections meeting in a point; the sound becoming condensed in this point as the common focus of all of them.

The vibrations in sonorous bodies may be stopped by the hand;

this effect we perceive in glass touched by the finger, when the vibrating motion immediately ceases, even if the glass be touched very lightly. Water conveys sound distinctly, and to a great distance. The surface being smooth, the incident sound is reflected from it, and, by the small angle it makes with the surface, is thrown to a great distance by repeated reflections in the same direction.

I cannot quit the subject of sound without speaking of its different states in the same latitude; which depend on the various circumstances of each portion of the earth, arising from soil and situation. We know that the highest elevation on the earth cannot be more affected by the heat of the sun than the lowest spot on its surface, because the distance of any part from that luminary is so great, that no sensible difference can be perceived by us, even were the heat principally derived directly from the sun, which is not the case.

The air being more dense in a valley and on a plain than at the top of a mountain, must consequently retain a greater quantity of the sun's rays; and by its receiving numerous reflected rays from the objects that surround it on all sides, it must necessarily be more heated.

The rarity of the atmosphere on the surface of a mountain, is according to its height; consequently its disposition to retain solar heat will be proportionably decreased, and its property of retaining the heat of the earth likewise diminished: this accounts for the extreme coldness perceived at the tops of high mountains, which are almost perpetually covered with snow, while the valleys below are in a high state of verdure and fertility. Travellers on the continent are well acquainted with these circumstances, and particularly with the different gradations of heat and cold perceived in ascending the Alps and Apennines. The different heights at which the extremes are evident depend, in some measure, on the latitude; for in the torrid zone these effects are not perceived but on the highest mountains; whereas in the temperate zones the regions of ice commence at a much less considerable elevation above the surface of the sea. In higher latitudes the elevation is still less considerable; and at the greater distance from the equator, in the frigid zones, even the lowest surface of the earth is perpetually covered with ice.

Many of the inhabitants of the Alps have glandular excrescences, called goitres, growing on their throats, which, from the size of a small nut, increase to a considerable magnitude. The cause of these tumours has been ascribed by some to the use of water produced by the dissolution of snow; by others to their drinking water impregnated with stony particles, as mentioned by Payne in his extracts.

The stagnation of the collections of water from which these people procured their unwholesome supply, perhaps contributed to render the air unfit for free respiration; for now that these sources, by the wisdom of their legislature, are dried up in many parts, and a supply of water procured from rivers, the number of goitres is much decreased: of course, those persons not previously afflicted, will not have them. Yet many of these people do not consider the goitre a misfortune, but rather an ornament; at least, they esteem them a distinguishing characteristic of their district.

We may suppose Europe to have been much colder in ancient times; for history informs us, that the rein deer, now the inhabitants of only the northern latitudes, Lapland and Siberia, on our continent, were formerly found in Germany and Poland. This difference of climate may be ascribed to the immense woods, formerly existing in these latter countries, obstructing the rays of the sun in their approach to the earth; which woods having been cut down, and the lands cultivated, these circumstances have produced the present temperature of the air in those parts. Canada, in North America, now represents what Germany was in former times. It has the same latitude with England, but rein deer are abundant in it, and snow constantly lies on the ground. From which we may conceive that the variations of heat and cold felt at different places having the same latitude, though in some degree dependent on the quantity of the sun's rays falling on each part, are principally occasioned by other circumstances, such as the cultivation of the land, &c.

Having explained all the mechanical properties of air, and noticed the amazing variety of effects produced by them, let us consider this element in its constitution and in its uses, and reflect on the wonder-working hand of Providence. The atmosphere is composed of all the active and volatile parts of animals, vegetables and minerals: it is capable of various modifications arising from heat and cold, moisture and dryness; transmits and reflects light; sustains water, and, by refrigeration, renders it the more fit for the use of man. This medium of existence, in its purest state, is a vivifying power, strengthening the organs of our bodies: but when obstructed, it renders us languid; or if surcharged with heterogeneous particles, it becomes no longer fit for the preservation of the bodies it was intended to sustain. As air acts on our animal nature, so does the pure religion taught by our blessed Saviour act on our spiritual part; for when our minds are unclouded by ignorance, and unobstructed by vice, it excites all our virtuous energies, animates us in the performance of our duty, and invigorates our love and adoration of the beneficence and profound wisdom of our great Creator. But how sad the reverse, if we suffer the contaminating principles of vice, so uncongenial to the true spirit of christianity, to possess our minds! By their pollutions our understanding becomes enfeebled, its vigour is impaired, and it no longer retains its active force; for the baneful influence of vice, like a gathering storm of the elements, will first obscure our reason, and then totally obstruct the light of revelation—that heavenly medium, which alone sheds beams of wholesome intelligence, to perfect the soul of man!

PLATE XII

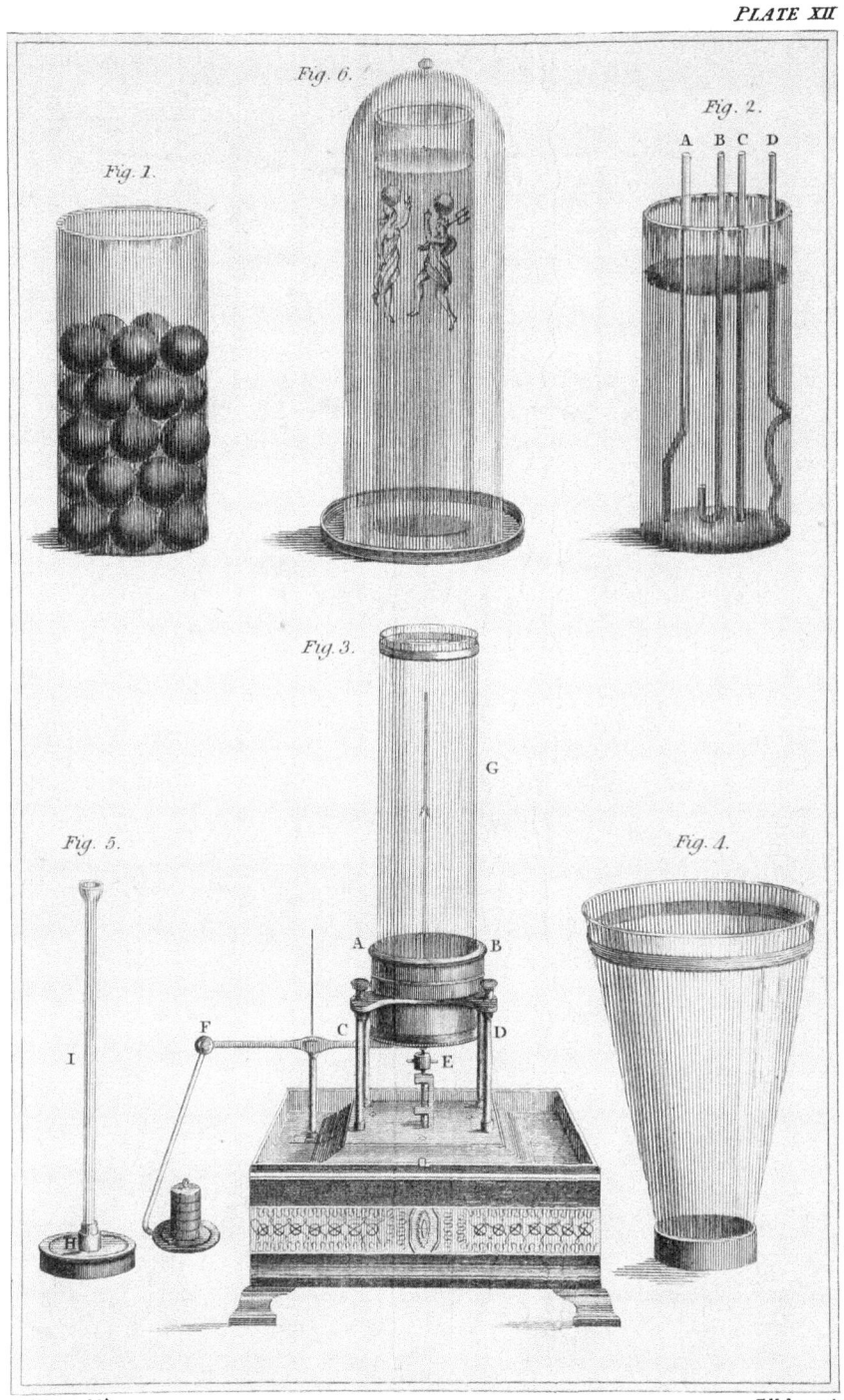

Fig. 6.
Fig. 1.
Fig. 2.
A B C D

Fig. 3.
G

Fig. 5.
Fig. 4.
A B
F
C D
I E

H

S.M.Bryan del.ᵗ   H.Mutlow sculp.ᵗ

PLATE XIII.

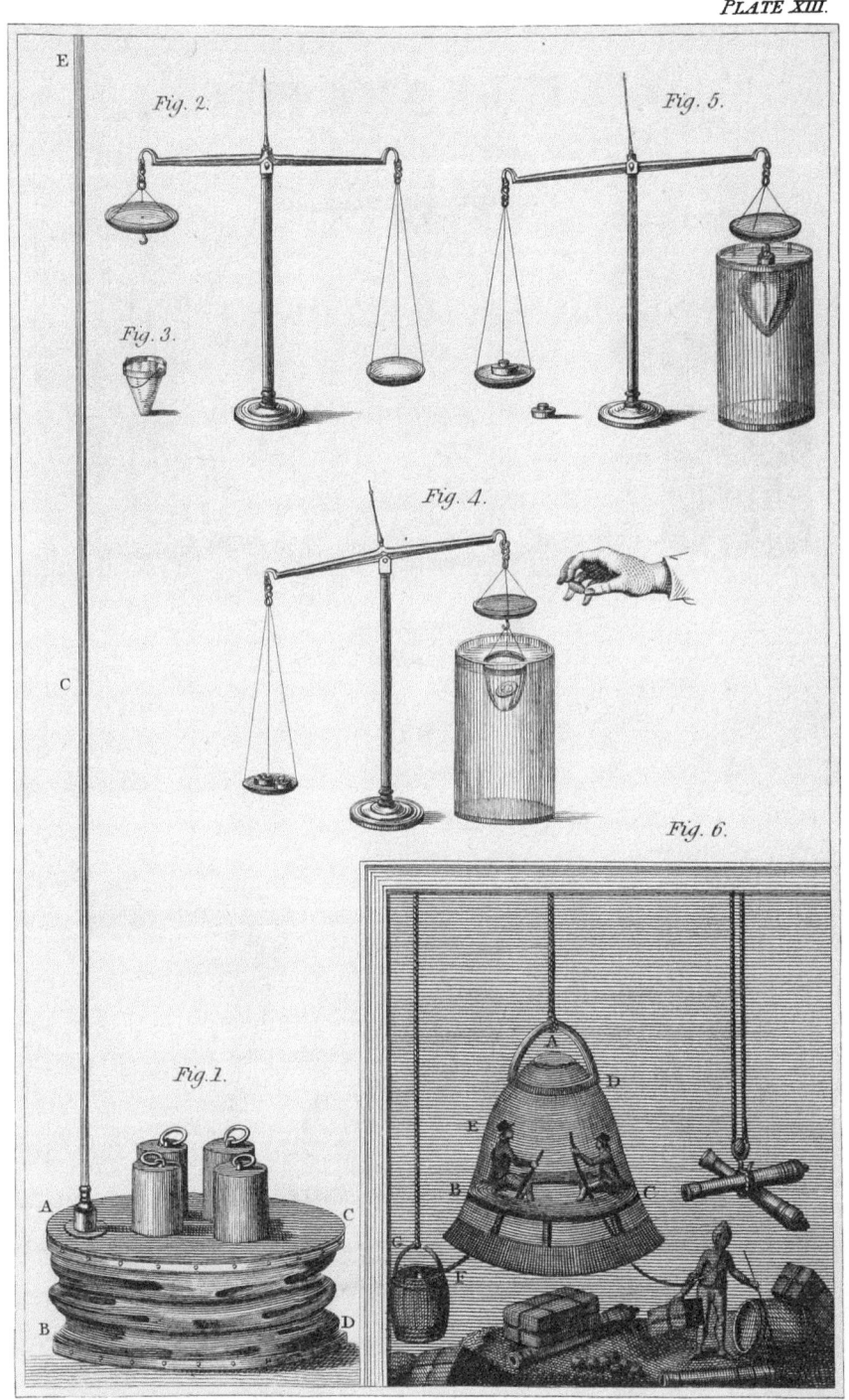

Fig. 2.

Fig. 3.

Fig. 5.

Fig. 4.

Fig. 6.

Fig. 1.

T. Noble delin.

H. Mutlow sculp.

LECTURE THE FIFTH.

ON WATER AND HYDROSTATICS.

Where lurk the vast eternal springs
That, like creating nature, lie conccal'd
From mortal eye; yet with their lavish stores
Refresh this globe, and all its joyful tribes.   THOMSON.

WATER in physiology is an element, inodorous, transparent, colour-less and tasteless; by cold condensed into solid masses, and by heat rarefied or expanded into a volatile vapour of amazing elas-ticity. In its intermediate or natural state, it is fluid, penetrating and non-elastic, and capable of opposing resistance to an impressed force.

Water is seldom free from heterogeneous particles, generally con-tains salt, and is impregnated with various other principles. The penetrative quality of this fluid is very remarkable; for the driest substances, even bones, which by time have become almost as hard as iron, by distillation have yielded half their weight in water. This water must have been imbibed from the atmosphere, which is con-stantly a recipient of the aqueous fluid. Water is both a dissolvant and a coagulator; for it separates the parts of some bodies by its

penetrative power, and mixes and binds others together by its cohesive quality; which, with other properties, render it of great and
extensive utility in all the various operations of nature and art.

The particles of water are moveable among each other, though
they adhere together with a force perceivable on their coming in
contact with another body.   Dip any substance in water, and there
will always be drops suspended from it, each drop composed of a
great number of minute particles of the fluid. This effect of cohesion
probably arises from the spherical form of the particles of water
which causes an equal action on their superficies, they mutually
acting on each other from their centre, and thus they form themselves into larger globules.

The agency of water may be considered as universal and effectual
as the element of fire, for it is an essential medium of all existence,
circulating with the blood of animals, and the juices of fruits, flowers
and vegetables.   To art it affords strength and facility of conveyance, connecting the trade and interests of mankind in distant
countries; so that it may be called the mother of commerce, the
universal harbinger of intelligence, the denizen of the world.

So equal are the uses and importance of this fluid in the natural
world, that in ancient times it was considered the first principle of
universal existence.   That it is a principle of life is certain, from its
universal diffusion in the earth and atmosphere: it mixes with the
nutritious part of all bodies, for which its insipidity, that negative
virtue, peculiarly qualifies it; for it does not alter the taste of substances by its impregnation.   Water purifies air, as is evidently
proved by passing polluted air through it, by which means the air

is rendered sweet. The more we examine the uses of water, the stronger evidence we shall perceive of the Wisdom that produced it, for such various and beneficial purposes.

Great quantities of this aqueous fluid being constantly supported in the atmosphere, and occasionally falling in showers, purifies the air by impregnation, and, by the motion it communicates, renders it fitter for the respiration of man and animals; while the impure particles falling to the earth, assist the indispensable process of vegetation. Of the existence of water in the atmosphere we have many familiar evidences; but the amazing quantity of that fluid constantly suspended in the air, can be understood only by experiment.

Dr. Watson, Bishop of Llandaff, endeavoured to estimate the exact quantity raised from a given part of the earth in a given time, by a very curious process. He exposed eight ounces of salt of tartar in the open air on a clear day, after previously drying it on a hot iron. In the space of three hours, from eleven to two o'clock, the salt increased in weight two ounces. In the course of a few days its weight was increased to twenty ounces; it was then quite fluid; and being distilled yielded nearly its increased weight of pure water.

It is very difficult to ascertain the purity of water. In general, the softest is the best. Snow water is very soft. Rain water is considered pure; yet we cannot entirely subscribe to that opinion, for it is usually impregnated with noxious particles from the atmosphere. The purity of spring water depends on its source and meandering progress; for, passing through different strata of the earth, it imbibes something of the quality of each; however, notwithstanding its appearance, it may be considered the least pure of any ; and therefore virtually, though not naturally, unfit for culinary purposes.

We learn by experience the nature of the water, and are able to determine which is beneficial or hurtful to us. But in regard to purity and perfection, the water that feels light on the stomach, boils vegetables quickly, and mixes with soap without curdling it, is reckoned the most pure and safe, as possessing no dubious property. Water is purified by distillation, which raises the pure parts, leaving the heterogenous ones at the bottom of the retort * Thus does the heat of the sun raise water in the atmosphere in the form of vapour, which, after remaining a certain time in that state, falls to the earth again as dew, rain, &c. washing from the atmosphere such particles as would by their corruption render it noxious and unwholesome.

> Thus, in his stupendous lab'ratory,
> The sun, arch chemist, unceasingly performs
> His daily task of subtlest distillation!

The combination of water with various noxious vapours must render the evening dew extremely insalubrious. The fluidity of water seems to require no explanation; yet as that characteristic is considered in its mechanical effects, I shall speak of it previously to the application of its properties, which will hereafter come under our consideration. We conceive a fluid to be composed of spherical

* A retort is a close vessel. One end of a long pipe is cemented in it, the other end communicating with a vessel below. The pure particles rising in steam, pass through the pipe into the lower vessel, which is kept constantly cold to regenerate the pure particles of vapour into water. I do not enter on the subject of chemistry in these lectures, as the experiments in that art are too inconvenient and dangerous for female performance and introspection.

particles, which form suffers them to move easily among themselves, and causes them to yield to every impression. And as a quantity of sugar or salt dissolved in water does not increase its bulk, we imagine that these substances mix with the particles of water, by filling up the interstices between the globules; as sand would those formed in a mass of spherical bodies, as exhibited, fig. 1, pl. 12, in which the bulk of the whole mass is not increased by the interstitial reception of the vagrant particles. The cohesive power of fluids is diminished by heat, and increased by cold; the former separating, and the latter condensing, their particles. The most solid bodies are fused by heat, and rendered fluid; from which we may infer, that the particles of all bodies are alike, their apparent difference arising from the various modifications, by heat or cold, moisture or dryness; and that the foundation of all bodies in masses, whether solid or fluid, consists of solid particles, and their associated density is in proportion to the quantity of ponderable matter they contain.

The solid particles of bodies, when in the state of fluidity, gravitate in all directions, and arrange themselves parallel to the horizon. The other laws which govern the motions of fluids we shall have occasion to contemplate when treating of Hydrostatics and Hydraulics.

Water in masses possessing the same gravitating tendency with solids, causes the water of the ocean to appear convex, when we look to sea; for its surface is arranged concentric with the earth, according to the universal law of gravity. This fluid is deemed incompressible, which may arise from the hardness of the particles and their non-elasticity. The following popular experiment on the incompressibility of water has been repeatedly performed by different persons, and the result has always been the same. A hollow copper

globe of considerable thickness being filled with water, and then
struck or compressed with great force, the water penetrates the substance of the globe, and appears on its surface.

We have before considered the elastic power of water, when
raised in vapour, otherwise the experiment I am going to relate
would astonish you: I conceive this expansion may arise solely from
the heated air in the interstices of the water, which separates its particles and propels them with force in the form of vapour, which,
when confined and violently heated, will burst the strongest vessel.
A small quantity of water being included in a globe of metal two
inches thick, and placed in a furnace; when the water becomes hot,
not being able to escape any other way, it will burst the globe in
pieces.

We may reasonably conclude that water is a fluid, from its
combination with fire in an active state; for when that fluid is
quiescent, the particles of water adhere together, no active force
inducing a separation of them; and supposing fire to reside with air
in the interstitial vacuities between the particles of water, yet when
inactive it does not increase its volume: from which also we infer,
that the elementary parts of water are solid.

A quick current of air facilitates the change of water into a solid
body: for water unagitated freezes slowly; but if exposed to a
stream of air, ice is very quickly formed; because the motion suffers
the element of fire to escape with the air, and thus the fluid is converted into a solid substance. When we expect liquor will freeze, we
should, to preserve the vessel that contains it, allow for its expansion
in the act of freezing; which is occasioned by the air confined in its
interstices endeavouring to escape, but not being able entirely to

effect that purpose, forming bubbles within the fluid, which extend its volume. That the bubbles of air cause the expansion of water in the act of freezing, has been proved by experiment on water deprived of air, when its volume was not increased by its change of state.

The ice of standing water is formed first on the surface; and gradually increased in thickness by one stratum of water freezing after another in quick succession. But the ice of flowing water is first formed in the middle, and increases gradually towards all parts of the surface of the stream. Frost generally proceeds from the upper parts of the earth downwards; but how deep it will penetrate is not known, as this must differ from a variety of causes, such as the duration of the frost, the texture of the ground, &c.

The frosts of a severe winter make dreadful havoc in the vegetable kingdom. They deprive the tender parts of trees and plants of their nutritious elements, by stopping the circulation of their juices: these effects we but seldom experience in any great degree in this climate; but the spring frosts, which are more frequent, are often fatal. Humidity is the intermediate agent of destruction, causing the greatest devastation among the vegetable tribe; for by its expansion in the act of freezing, it bursts the tender buds and shoots, and thus destroys their vigour and vitality; and sometimes even the solid trunks of large trees are destroyed in this manner. These effects of humidity are particularly evident in the shoots of flowers and plants, when a frost follows heavy dews, or a long season of rain. Hence, by deduction, we comprehend why plants that grow by the side of a river are often destroyed by the spring and autumnal frosts, while those of the same species which stand in a drier place survive the torpid influence; for though their vigour is suppressed for a time, yet it is seldom annihilated. I think there is no doubt that the

P

failure of vegetables, and some species of flowers, by the sea-side, is occasioned by the humidity of the atmosphere. The air, being filled with saline particles, is continually moist, which occasions the frost to act more powerfully on vegetation.

A full theoretical investigation of mineral waters is here un-necessary; I shall therefore only briefly explain the meaning of the term mineral, which is applied to water much impregnated with foreign principles. We find that our ancestors were particularly attentive to the properties of water, at least to its effects on the con-stitution of man; for it was not its abundance alone that deter-mined them where to dwell, but also its quality that influenced their choice of situation.

Hippocrates, the father of medicine, was well acquainted with the effects that water has on the human constitution, and found that its analysis was essentially necessary, in order to determine whether any particular water was useful or hurtful, and if it possessed any medicinal virtues; and by his application to this subject, he per-ceived that mineral waters possessed qualities adapted to the different constitutions and disorders of mankind. Water imbibing the minute particles of substances it meets with, becomes possessed of their different qualities.

Experiments.

Experiments will show us what substances water has passed over. If it have imbibed an alkaline quality, the syrup of violets turns it green; if it be charged with acid particles, the same syrup makes it red: when it has run over iron ore, a solution of galls turns it black; and when it is impregnated with alum, oil of tartar renders it turbid. If either of the above-mentioned qualities be obnoxious to the constitution of individuals, they may thus by experiment avoid the danger. Water impregnated with lime-stone, and oozing slowly among mosses, leaves, &c. adheres to those substances; when, evaporating, it leaves the stony particles in them: which curious result produces the beautiful petrifactions we meet with in Derbyshire, Yorkshire, &c.

Two favourite theories have long maintained their sway in the minds of philosophers; of which were I to venture an opinion, it would be, that both are right in some particulars, and wrong in others; but I will relate them, and leave the absolute reception or rejection to abler judgment.

The water raised in vapour from the ocean, has been found by calculation sufficient to supply the quantity furnished by all the rivers on the earth; and it is thought probable that the sea receives from the rivers what it loses by evaporation, this fluid thus circulating through them, and through the air of our atmosphere. The vapours that are raised from the sea, being carried by the wind, and condensed against the sides of mountains and other eminences, where, trickling down through the crannies, they enter the hollow parts of the earth, form collections of water, which issuing out at orifices and falling into low places, constitute springs. The other theory of springs and rivers refers their origin and support to a great abyss of water occupying the central parts of our globe, all the phenomena of springs being chiefly derived from the vapours and sluices of this great body of water into which they are all returned; and that thus a perpetual circulation and equality are kept up; the springs never failing, and the sea, by reason of its communication with the subterraneous water, never overflowing. Now I conceive it probable, that the source of springs and rivers may appear to be different, and yet both originate in the same cause.

Suppose we admit that some rivers, springs and fountains, derive their source directly from the evaporation of the sea, attracted by the sun, and conveyed by the atmosphere to their repositories; yet may there not also be an internal supply by an accumulation of this water within our globe? This will equally explain the circumstances advanced to authenticate the former theory of those springs which appear independent of external circumstances, and which are uniformly the same in rain or drought, for they certainly do rise from an internal supply: and, as a further evidence in support of that opinion, wherever we dig deep enough, springs are always found. In confirmation of the above, we may also infer that if the internal sources of them were from the ocean, through fissures, they would rise to the surface of our earth; which is not the case, for they have different levels; yet may these effects arise from collections of water formed in the earth, which water was originally derived from the atmosphere of the earth in the manner just described. On this subject, embarrassed with equivocal and contradictory hypotheses, I shall not attempt to be oraculous; but merely observe, that water from the centre of the earth could never rise to the surface of it on any known hydraulic principles.

Of the cause which produces the saltness which characterizes sea water, no satisfactory natural solution can be given; nor does it seem necessary, for one perfectly consonant to reason and religion presents itself—that the Almighty endowed this large body of water with this property for the preservation of itself and the creatures which inhabit it.

ON HYDROSTATICS.

The science of Hydrostatics is properly the estimation of the specific gravity of bodies, both solid and fluid; and the weight, pressure and action of fluids in general, on solids and on each other, either at rest or in motion. These affections have been investigated principally in regard to water. The term Hydrostatics, when used in its most extended sense, includes the science of Hydraulics, which describes and calculates the motion of water through pipes, &c. But I shall avoid associating the two branches, to prevent confusion in the distribution of the properties and affections belonging separately to each of them.

The various affections of fluids are still perplexed with difficulties not resolvable on mathematical principles; yet we know enough of them to render the application of our knowledge extensively useful.

Experiment.

To prove that the gravitation of fluids is in all directions, and with equal force in each of them, we will take the open glass tubes, A, B, C, D, fig. 2, pl. 12, of different forms, and immerse them in water; when we shall see the fluid rise in all of them to the same level, and at the same instant of time. If we place a finger on the top of each glass and remove them altogether, the effect will be more striking. On immersing the tubes, the air in them prevents the fluid rising more than a quarter of an inch; for these being full of air, the water cannot displace it altogether, though it compresses it a little. When the air is suffered to escape at the top of the tube, the water, gravitating in all directions, rises to the same elevation in each tube, and to the height of the surface of the surrounding fluid; yet it enters with a different direction in each, namely, obliquely, horizontally, and perpendicularly, upwards and downwards. The water rising in each to the height of the surrounding fluid, shows that the fluid endeavours to place itself on a level with its source.

The surface of a fluid that is free from all impediments is parallel with the horizon. Therefore from this tendency, and the laws of its gravity, the pressure of a fluid on the bottom of a vessel is in proportion to its height and base.

Experiment.

Let fig. 3, pl. 12, represent a machine to measure the pressure of water; A B C D, a brass cylinder which forms a base to each of the glass appendages. The bottom, E, is supported by a weight suspended from the lever F, which keeps the cylinder water tight. If the inverted glass cone, fig. 4, the smaller cylinder, fig. 5, and the larger one, G, be filled with water to the same height, we shall find that the fluid exerts an equal force against the bottom of all of them; though the cone, fig. 4, and cylinder G, fig. 3, contain each a much greater quantity of water than the cylinder, fig. 5. This we prove by pouring water into the cylinders, when the weight on the lever will be overcome at the same height of the water in each of them.

The bottoms of the glass vessels have brass rims, which screw on to the cylinder, A B C D, of the machine, so that they all have the same base. The base of the small cylinder, fig. 5, has a top, H; when water is poured into the glass tube (as water gravitates in all directions, and endeavours to rise to the same level) it exerts a force against this top, by endeavouring to rise to a level with the water in the cylinder; this causes the brass top to oppose an adequate reaction on the water, which makes that exert the same force against the bottom, E, fig. 3; which force is equal to what it would be if the cylinder, fig. 5, were of the same size with the cylinder G, fig. 3. In regard to the cone, fig. 4, though it contain more water than either of the cylinders, yet the pressure of the fluid on its oblique sides causing a reaction, the superabundant weight is supported; hence, the pressure of that quantity of water when placed on the bottom C D is no more than in the other vessels.

Experiment.

The hydrostatic bellows serve to authenticate the experiment of the hydrostatic paradox above related. Fig. 1, pl. 13, represents this machine. A, B, are two oval boards; which open and shut to a certain distance. These bellows have not valves, like the common ones. The pipe c is fourteen feet high; water poured into it runs through into the bellows, and raises the upper board A. On filling the pipe and the cavity between the boards with water, the fluid raises and supports as much weight laid on the upper board, without pressing the water out at the pipe, as is equal to what a whole cylinder of water would weigh, having A B C D for its base, and of an equal size all the way to the top of the pipe at E, which would be several hundred weight. For, suppose the pipe to be fourteen feet high, the boards sixteen inches in diameter; square the number, multiplying 16 by itself, and the product is 256. For a circle we subtract one-fourth, which leaves 192 inches for the area, which multiplied by 14 feet, makes 2,688. There are 144 inches in a square foot, and 18 times 144 in 2,688. A cubical foot of water weighs 1000 ounces; and, as there are 16 ounces in a pound, if we divide 1000 ounces by 16, it will go 62 and a half times; 18 times 62 and a half pounds is 1,125 pounds for the whole column of water; and therefore that weight may be laid on the upper board of the

bellows before the water will run out at the top of the pipe ; the reaction of the upper board being equal to the endeavour of the water
to rise to its level in the pipe: for the pressure of the water between
the boards must balance the water in the pipe, or it would run over;
which effect the upward pressure on the board prevents.

Archimedes, a famous geometrician of Syracuse, a city of Sicily,
fully investigated the properties of water, and the floating of bodies;
their relative gravities, levities, situations and positions. He was the
first man who attempted to determine in what proportion bodies
differ from one another as to their specific gravities; and by this investigation he discovered the cheat of the workmen who had debased
king Hiero's crown.

In order to comprehend the manner of ascertaining the specific
gravity or weight of bodies, we must first understand that the term
density means the proportionate quantity of matter contained under
a given surface, which is estimated by the ratio its weight bears to
its bulk; for the more numerous the particles in proportion to the
space they occupy, the greater is the weight or density of the body.

The different weight of bodies of the same size is called their
specific gravity, being their comparative weight. The mode of
estimating the specific gravity of bodies is through the medium of
common water, in which they are immersed for that purpose *. On
the specific gravity of bodies the whole rationale of sinking and floating depends. All tangible fluids are nearly incompressible, and
experiments have proved that even water is nearly so ; yet the experiment I have related of a globe filled with water is not quite so

---

* See table of specific gravities, Appendix.

satisfactory a proof of this assertion as I could wish; for the penetrative power of water renders it less decisive in determining its
degree of incompressibility: but a clearer deduction presents itself
in the following experiment.

Canton took a glass vessel, large at bottom, and terminating in
a small tube at top, and, filling it to the brim, he tied a bladder over
it; on pressing it down, the bladder displaced but a very small portion of the fluid: indeed I think the depression might not be in the
least occasioned by the elasticity of the water, but merely by the
condensation of the air in the water.

On the force water exerts against an impressed power depends
the rationale of floating bodies on the surface of it; and as all solid
substances, by the force of gravity, tend downwards, it depends on
the absolute weight of an immersed body whether it shall sink or
swim.

If the weight of the body be greater than that of an equal bulk
of the fluid in which it is immersed, the excess of force downwards
will cause the body to sink. Should the weight of the immersed
body be less than that of an equal bulk of the fluid, the upward
pressure will prevail, and cause the body to float on the surface. If
both are precisely of the same specific gravity, the body will remain
at rest in any part of the fluid. Thus, a stone being specifically
heavier than water, when thrown into that fluid, it sinks by the force
it exerts to descend, being greater than that of the water in opposing
its descent.

When a body is specifically lighter than the fluid in which it is
immersed, it rises, and floats on the surface; for on placing the body

Q

in the water, it is pressed on all sides equally by the fluid : yet the immersed body displaces no more of the fluid than is equal to its own weight ; therefore if the body be specifically lighter than the fluid, this exerts its remaining strength, or difference of weight, upwards against the body, and supports it, in proportion to the difference of their specific gravity.

It is on this principle that vessels float on the surface of water; for though the materials are specifically heavier than that fluid, yet, being hollow in some parts, they may be altogether, bulk for bulk, specifically lighter.

Let us advert to the wise allotment of power in fishes, to enable them to regulate their motions in their natural element. They are furnished with an air-bladder, by the expansion or contraction of which, they rise higher, sink lower, or swim on the surface of the water. The usefulness of this power may be understood by the observations made by the honourable Robert Boyle, recorded in the Philosophical Transactions. Observing a bubble of air rise in the water, he perceived that it expanded itself as it rose towards the top. This he justly ascribed to the different pressure of the water on it, at different depths in that fluid. Thus, we conceive the air-bladder in a fish counterbalances its different situations in the water, by regulating its specific gravity ; for, being a part of the fish, when it is collapsed the size of the fish is diminished, and becomes specifically heavier than the water it inhabits: hence, when the air-bladder is greatly expanded, the fish is specifically lighter than the fluid medium; and at other times of the same specific gravity with it.

Though the cohesion of the particles of water seems to make no part of the rationale of large bodies floating on its surface, yet very small particles, even of gold and iron, though specifically much heavier than water, when they are finely pulverised, float on the fluid, for their weight is not sufficient to overcome its cohesive property.

The part of bodies which sinks below the surface of the surrounding fluid, is equal in bulk to a quantity of the fluid that weighs as much as the whole body; for which reason a body having the same specific gravity with the fluid in which it is immersed, rests in any part of it where it happens to be placed; for equal masses balancing each other, the force with which the body endeavours to descend is exactly equal and contrary to the force the water exerts in opposing its descent, and consequently it remains at rest.

Many important uses result from the knowledge of the specific gravity of bodies: to the further illustration of this subject we will now proceed; first recurring to the foregoing observation, that the pressure of a fluid upwards lessens the gravity of an immersed body. Thus, any substance immersed in water seems not so heavy as when out of it; for the upward pressure supports a part of its weight, and is equal to the difference of their specific gravity. This truth, that the part or the whole of a body below the surface of water loses as much of its weight as is equal to a quantity of the fluid of its own bulk, is the foundation of the principles demonstrated by the hydrostatic balance, represented fig. 2, pl. 13; which is a very nice scale-beam, to which a short and a long scale are suspended, balancing each other.

Hydrostatic Experiments.

To weigh a substance, we place it in one scale, and its balance weight in the other. Fig. 3, is a bucket used for the purpose of weighing bodies in water. Suppose, for instance, a guinea be weighed in the air, and the bucket, fig. 3, filled with water, and having its balance weight in the other scale. On immersing it, as seen at fig. 4, and placing the balance weights—of the guinea out, and of the bucket in water—in the longer scale, we shall find it necessary to put a weight in the water scale, to keep the beam in equilibrium; because the water supports, by its reaction, as much of the weight of the guinea as a quantity of water of the same bulk with itself weighs. To ascertain the specific gravity of a guinea, we divide the weight of it in air by its weight when put into the water scale; and if it go seventeen times and three-quarters, the guinea is sterling. This hydrostatic experiment is the only true method of discovering alloy in gold; for, knowing what a certain quantity of gold should weigh, its deficiency will always be ascertained by this means. A counterfeit guinea may have the standard weight by its size being increased; but this will be discovered by its loss in water.

That the weight lost by a body immersed in water is always equal to what a quantity of the fluid of its own size weighs, is evident by the experiment of weighing a piece of lead, exactly of the same size with the bucket, in air; filling the bucket with water, and, when all is in equilibrium, immersing the lead in a receiver of that fluid, when the bucket will preponderate; but on emptying the bucket of water, the balance will be restored. The same weight would be lost were the immersed body wood, or any other substance; it being in proportion to the bulk, not to the density, of the immersed body.

To discover the specific gravity of a substance lighter than water, a piece of cork for instance, first weigh it in air, and it will be found to require more than four times the weight to sink it in water that was necessary to balance it in air ; therefore the cork must be more than four times lighter than water. Understanding that a body immersed in a fluid loses as much of its weight as is equal to its bulk of the fluid ; to ascertain the specific gravity of other fluids, we need only immerse the same solid in each of them, and observe the difference in the loss of weight in them and in water.

Fig. 5, pl. 13, represents a conical piece of glass suspended from the water scale, the specific gravity of which we wish to ascertain by immersing it in the fluid, with its balance weight in air in the longer scale. It now requires additional weight in the water scale, to restore the equilibrium. Thus, by comparing its specific gravity with that of water, as a standard, we may find the difference in the weight of any other fluid. If the fluid be lighter than water, the solid will sink lower in it ; if heavier, it will rise higher than in water. Hence, in the former case more weight will be necessary in the longer scale, and in the latter less, than when the solid is immersed in water. From these experiments we learn that the difference of weight proceeds from part of the weight of an immersed body being sustained by the fluid ; and which is the reason that the whole weight of a bucket of water is not felt while it is below the surface of the water in the well. As the gravity a body loses when immersed in a fluid is always the weight of as much of that fluid as is equal in bulk to itself, the weight lost by the body cannot depend either on the depth of the fluid itself, or the depth to which it is immersed.

Hence it is not more easy to swim in deep than in shallow water, provided there is depth sufficient to permit the act of striking out the limbs freely ; for, whatever be the depth of the water, a man is supported exactly in the same degree, and in proportion to the difference of specific gravity between himself and the fluid. It is easier to swim in the sea than in rivers, because salt water is specifically heavier than fresh; and as a man immersed in salt water is supported by a quantity equal to the bulk of his body, and in a river is supported by a relative quantity of fresh water, the power gained in salt water buoys him up with superior force.

It is supposed that there are very few, if any, animals whose specific gravity is greater than common water, though the solid parts of animals are heavier than that fluid. Their floating on the surface of water arises from the air, oil, &c. contained in the cells or receptacles interspersed within their bodies, which substances are lighter than water, so that, taken together, they form a mass specifically lighter than common water.

I cannot conclude this part of the subject without introducing the ingenious contrivance of the Abbé de la Chapelle, who, in 1767, invented a jacket, capable of supporting a man in water. The experiment was made by himself. He contrived this dress on the known principle of specific gravity. Between two pieces of cloth, formed into a jacket without sleeves, he quilted pieces of cork an inch and half square, and three-quarters of an inch thick, close enough together to admit of a certain quantity being introduced into the given space, but leaving room for flexibility. This jacket is fastened on by strong buttons, and has a girdle to keep it steadily in its proper situation. By this contrivance the body is supported in an erect posture, without any effort on the part of the wearer, who rests his arms on the surface of the water. The utility of this invention being ascertained, I think, among other great patriotic exertions, it would be highly useful to the state, and honourable to humanity, if every vessel fitted for sea were furnished with a number of these jackets, to serve in times of danger and necessity. An attention to the claims of humanity, that godlike virtue, in all public transactions, serves to stamp the character of a state and people, and to raise them in the estimation of other nations. Universal philanthropy being the genuine and distinguishing characteristic of true nobleness of nature, and correctness of principle.

PLATE XIV.

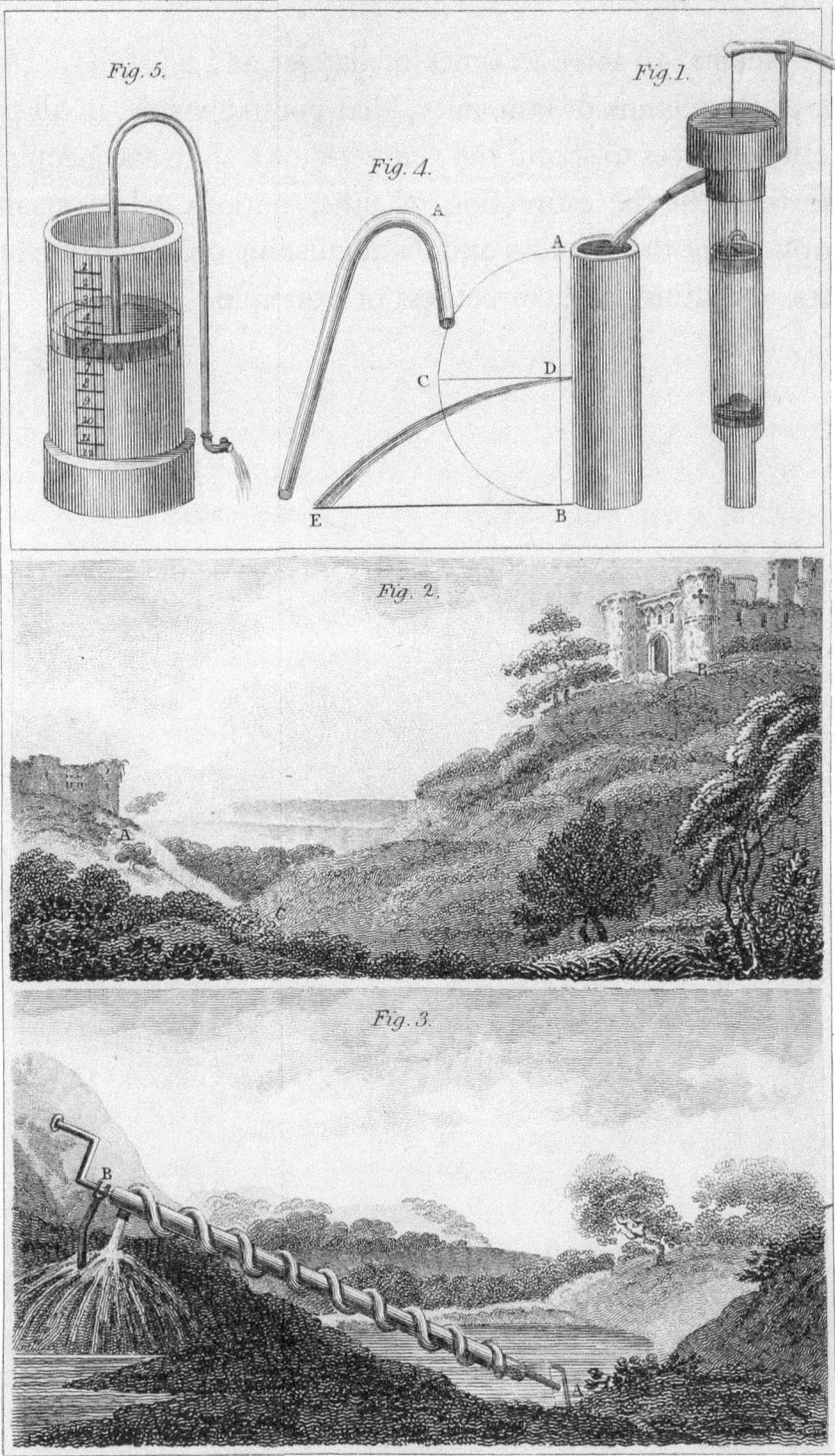

Fig. 5.

Fig. 4.

Fig. 1.

Fig. 2.

Fig. 3.

M.Bryan design.   S.M.Bryan delin.   H.Mutlow sculp.

PLATE XV.

Fig. 1.

Fig. 2.

Fig. 3.

M. Bryan design.   S. M. Bryan delin.t   H. Mutlow sculp.

LECTURE THE SIXTH.

A CONTINUATION OF HYDROSTATIC PRINCIPLES, AS EMPLOYED IN AEROSTATION, &c. ALSO ON HYDRAULICS.

" God never meant that man should scale the heavens
" By strides of human wisdom."   COWPER.

THE evidences of the principles of hydrostatics have produced many contrivances; for as the specific gravity of a body may be varied so as to be at one time greater, at another time less, than the fluid in which it is immersed, and at another equal to that of the fluid, the body either sinks, rises, or remains suspended in it, according to these varieties: on these known principles the modern practice of raising balloons, and adventurers in them, is founded. Aerostation is a curious experiment, but by no means subservient to a better purpose; except to that of convincing man how vain are his endeavours to counteract the dispensations of nature. Till the time of Bishop Wilkins, in 1672, aerostation was not thought of in this country. The act of flying was attempted so early as the time of Archytas, 408 years before the Christian era; but his flying pigeon we may suppose was constructed on mechanical principles, as he was a most ingenious mechanist.

R

Friar Bacon had written on the subject of aërostation so early as the year 1292; but the properties of inflammable air not having been ascertained at that time, his opinion, though ingenious, probably was not verified by experiment. He conceived also that, by means of artificial wings fixed to the arms and legs, a man might fly as well as a bird; but, on trial, the power of the muscles which move the arms was found inadequate to the performance, being insufficient to strike the air with the force necessary to raise a man from the ground. Other circumstances would also prevent the flight of the human body; such as its form, and the organisation of its respiring functions; all creatures being fitted by God for their different stations, and appropriate performances.

In the year 1672, Bishop Wilkins affirmed the air to be navigable; on which static principle he fallaciously supposed, that any metallic vessel being filled with rarefied air—a lighter fluid than the elementary air—would be buoyed up in the atmosphere. He wrote a treatise on this subject, but did not pursue the idea, because we may suppose, on a fuller investigation, and taking the incidental circumstances into consideration, he discovered the inefficacy of the means to effect the scheme proposed. Francis Lana, the jesuit, contemporary with Bishop Wilkins, impressed with similar ideas on the possibility of navigating the air as communicated by the Bishop in his treatise, practically pursued them, conceiving a globular vessel—which form is more capacious than any other—when exhausted of its air, would weigh less, and might be made so large, as not only to raise itself in the air, but carry up passengers with it. His theory was unexceptionable, yet the means proposed were insufficient to accomplish the end desired; for, on experiment, he found that a vessel of copper so large as to float in the atmosphere was utterly incapable of resisting the external pressure; and thus his specious speculation terminated in disappointment. In the year 1766, Mr. Henry Cavendish ascertained the weight and other properties of inflammable air, determining it to be at least seven times lighter than common air; which discovery was the foundation of the present principle of aerostation.

It is unnecessary to enumerate all the persons who have made aerostation their study, as the retrospection would afford us no solid satisfaction; though the result of all their experiments may serve to shew the insufficiency of science to enable us to soar into regions not intended for our penetration. It is curiosity that prompts the desire, and the ebullition of genius that attempts the performance which is only for the prize of folly; for could we arrive at the greatest perfection in this art, most probably it would end in our destruction. Therefore on this subject suffice it to say, that the first aerial voyage was made in France, the adventurers in which were Mr. Pilatre de Rosier and the Marquis d'Arlandes. The balloon that conveyed them was supported by the hot air of a fire: they experienced no inconvenience in their voyage, which lasted twenty-five minutes, during which time they passed over a space of about five miles. They met with several different currents of air, and the balloon was in danger from the fire by which it was sustained. The liability to accident from this mode of filling the balloon, induced a trial of inflammable air; which, by its less specific gravity, was more manageable and fitter for the performance of longer voyages, not requiring the supply and burthen of fuel.

Many attempts have been made to give balloons a progressive motion, but without effect, and for the following reasons. The extensive surface the balloon opposes to the resisting medium through which it passes, occasions it to displace such a large quantity of air, that it cannot be moved with any degree of velocity in a horizontal direction: for, supposing the voyagers to give it a small degree of velocity for a short time, the resistance of the air increasing with the velocity, their strength must be employed in overcoming that resistance, which of course must diminish the velocity. Therefore, though balloons do move horizontally, that motion is occasioned by the current of air, for it is probable they cannot be steered in that direction by human means, to any great distance. As balloons are raised on hydrostatic principles, and such experiments are familiar, I thought it not improper to notice the subject; and though I am not fond of dwelling on matters merely speculative, these not being calculated to interest attention or excite emulation, yet I cannot refrain from illustrating the principle of aerostation by one or two simple and pleasing experiments, that you may perfectly understand the nàture of bodies rising in the air—a curious branch of hydrostatic knowledge.

Experiment.

Rolling this sheet of paper in a conical form, and securing the top with a pin, we will hang it to the side of the hydrostatic balance, and place it in equilibrium by a weight in the other scale; when, on rarefying the air within the cone by the flame of a candle, the cone will rise, and require considerable weight put into the scale to which it is attached to restore the balance.

Experiment.

In this jar of water are two hollow glass images, each with a globe or balloon over its head. These contain just as much water as renders the images specifically heavier than that fluid. After the air is sufficiently exhausted by the action of the air-pump, that within the balloons expanding itself, forces out the water, and the images rise to the top of the receiver. When the figures and balloons are specifically lighter than the water, they ascend to the top of the vessel; but if we stop the action of the pump at the instant they have acquired the same specific gravity as the water, they will remain suspended in any part of the fluid. When we restore the air to the receiver, the water rushes into the balloons, and the figures fall to their original station.

This experiment fully illustrates the theory of balloons, and other bodies, rising when specifically lighter than the fluid in which they are placed; sinking in it when specifically heavier; and remaining in any part of the fluid when they have exactly the same specific gravity: on which simple principles the whole process of aerostation depends. And it is the same thing, whether we render the air lighter by heat, or inclose a quantity specifically lighter than the common atmospheric air in a certain space; in either case will the body ascend till it arrive at that part where the atmosphere is of the same specific gravity with itself; and there it will float, but cannot rise higher.

We will now proceed to contemplate the most unequivocal and useful effort of genius in the science of hydrostatics—the art of diving in water. The diving-bell is a machine both of extensive use and scientific construction. It is usually formed like a bell; and, being full of air in its natural state, is suspended from a ship, and let down into the water. Air being a compressible fluid, the water rises in the bell only to a certain height. As air is a substance, it cannot be entirely excluded from the machine, and therefore occupies the upper part, and by the compression is rendered more dense, which causes it at first to exert a painful pressure on the divers; but after a little time, the air in the pores and cavities of their bodies becoming of the same density, the uneasiness ceases.

Doctor Halley, in a diving-bell of his own contrivance, remained for the space of an hour and a half at the bottom of the sea, at the depth of fifty-two feet. This bell, made of copper, was three feet in diameter at the top, five feet at the bottom, and eight feet high; it was loaded with lead at the bottom, so as to sink it steadily and gradually in the water. It had a valve in the side, and another at the top; the latter served to let out the foul and heated air, while the other received fresh air from two barrels which alternately supplied the deficiency of the vital principle, and expelled the foul air; by which means the divers, who received light from a large strong glass at top, could remain at the bottom of the sea, in order to search for goods, or fix cords to wrecks of ships, and for such-like purposes.

The water being prevented, by the air in the diving-bell, from rising in any considerable degree, the diver is able to breathe and live for some time. Keeping his head above the water in the upper part, he respires the condensed air; when he descends below the bell, he has a flexible tube communicating with it, through which he breathes. By the two barrels of fresh air, continually and alternately let down to the divers, the supply is communicated in such plenty, that Dr. Halley informs us, that he was one of five who were together at the bottom of the sea, in ten fathoms water, for above an hour and a half, without experiencing the least inconvenience; and that he might have continued there as long as he pleased, for any thing that appeared to the contrary. So much light was transmitted by the glass at the top of the bell when the sun shone, and the sea was calm, that he could see perfectly well to read and write, and to find any thing that lay at the bottom; but in dark weather, and when the sea was rough, nothing could be distinguished: this inconvenience he remedied by burning a candle in the bell, which he found vitiated the air in the same proportion a man does in respiring; both requiring an equal quantity of fresh air for their support, which is to the amount of nearly a gallon in a minute.

The art of diving is now so much improved, that one of the divers may be detached to the distance of a hundred yards, having a close cap put over his head, with a glass in the fore part, to see through, and a pipe communicating with the bell to supply him with fresh air. The chief consideration being to furnish the divers with fresh air, the air vessels are so contrived as to rise and fall on hydrostatic principles; as thus, when the air leaves one barrel, that rises, and the other falls.

In the first application of the diving-bell by Dr. Halley, the supply of air depended on signal from the bell, which mode was attended with danger to the divers, as once happened; when on sinking the bell, the signal ropes twisting with the other ropes, their play was prevented; and thus, melancholy to relate, two men lost their lives. When the water is muddy, the light appears red on bodies below its surface; this arises from the red rays of light having more force, and therefore penetrating heterogeneous bodies that are not quite opaque. We perceive this effect in the setting sun: the air on the surface of the earth, being moister and more obscure than that above it, causes the red rays to penetrate after the others cease to appear in that part of the atmosphere.

Fig. 6, pl. 13, represents the diving-bell constructed by Dr. Halley. A is the strong convex glass, to give light to the bell; B C is a circular seat for the divers: at D is a valve inside the bell to let out the contaminated air, and at E is one to admit the fresh air, or else a flexible tube introduced into it through the water, as at F.

G is one of the two casks of air, each containing thirty-six gallons. They have each a hole in the bottom, to let in the water as they descend, by which the air is forced into the bell through a flexible tube from the top of the barrel. The barrels are loaded, to make them steady and certain in their operations.

The art of diving has been long practised in the Mediterranean, and other parts, to procure pearls, corals, sponges, &c. But the method employed—namely, wet sponges, coated with oil to keep in the air, held in the mouth—can support life for only a very short time.

The importance of hydrostatic knowledge needs no particular eulogy; it is sufficiently acknowledged by its extensive usefulness; and of its affections we know enough to prevent great inaccuracy in the application of its laws, and to render them of extensive utility in the affairs of life.

As our spirit, our understanding, rises specifically above the gross materials of our corporeal frame, so let our actions bespeak that specific virtue, and raise us in the estimation of the world, the affection of our friends, and, above all, by the specific power of a good conscience, elevate us above the fatal effects of human occurrences, and direct our flight to a still more exalted station in the regions of bliss.

HYDRAULICS.

THE science of hydraulics relates to the mechanical and natural motions of water, in pipes, conduits, &c. The laws which govern the flowing of fluids are not in all cases perfectly ascertained; for neither the exact time a certain quantity of fluid flows from a given aperture, nor its degree of velocity, is accurately known; therefore our knowledge of hydraulics is circumscribed within narrower limits than that of hydrostatics. We cannot estimate the exact velocity of the current of a river, nor trace its path, so as to keep it within certain boundaries; neither can we calculate how the waters of a river would be raised by a lock, nor assign the best possible shape for the mouths of canals and water-courses, nor determine the exact form for ships and boats, that they may move with the least effort and resistance; nor calculate the precise force necessary to cause a body floating on the water to proceed with a certain velocity, nor yet determine the exact resistance of water to all surfaces:—yet we know enough of each of these, to render our knowledge profitable to us; and our calculations generally enable us to produce the effects essential to our purposes.

Water is of very great importance in the mechanic arts; mills, engines, pumps, &c. being all formed on a due estimation of its affections, from which we derive the principles that enable us to construct and work them with the greatest possible advantage.

On the Spouting of Fluids.

A vessel of liquor, having an aperture at bottom, with a plug in it; if we withdraw the plug, the fluid at first flows out with more force, and greater velocity, than when the height of the liquor is diminished. If the hole be in the side of the vessel, the effect is the same. The distance to which water will spout from a given place has been ascertained on the principle of the gravitating and pro- jectile forces, and may be understood from fig. 1, pl. 14; which represents a vessel kept full of water by means of a pump, the sur- face of the fluid constantly remaining at the same elevation, and

s

flowing through a horizontal spout at D, placed exactly in the centre of its height. Describe a semi-circle A B C, with the radiis D C, which being drawn from its centre, and perpendicularly to the line A B, is the largest that can be described from the diameter of the circle to any part of its circumference. While the vessel is kept full, the water will continue to spout from D, to the horizontal distance E B, which is double the height of the fluid, and also of the perpendicular A B. At whatever part of the side of the vessel the spout is placed, the distance to which the water will be projected may be calculated in the same way, by taking the perpendicular or sine of the arch at the height of the spout, and doubling it.

The known affections of flowing water are of great use; as may be perceived by supposing a reservoir of water at fig. 2, pl. 14, a certain quantity of which must be conducted to A in a given time. To ascertain what quantity of water will be discharged in a certain time at that place, it is only necessary to find the height of its descent from B to C; because when it arrives at C, it will have gained so much force as would carry it to the same height as from that it descended; but discharging itself at A, it will do so with a power proportioned to the velocity that remains of the force gained in its descent. The fall of water in an oblique or a perpendicular direction being always in proportion to the square root of the height of the fluid above the aperture through which it descends; estimated according to the time a body would fall that distance, which is computed to be in proportion to the square root of the height of the fluid; its force may always be calculated with accuracy, and also the quantity it will discharge in a given time at a given place.

As Archimedes' screw is often adverted to, I will endeavour to explain its hydraulic uses. It consists of a long cylinder, round which a spiral tube is twisted, and is worked by turning the handle. (Vide fig. 3, pl. 14.) This machine being placed in water, in an oblique position, the fluid enters the lower orifice A, and continues rising by the revolving motion of the cylinder, and discharges itself at B.

The conveyance of water by means of pipes to a great distance, has saved much labour and inconvenience; for till the universal tendency of the gravitation of fluids was known, large aqueducts, on immense arches, of extraordinary height and stability, were formed at an incredible expence, to conduct water from one place to another; which purposes are now effected with great ease, and at a comparatively trifling expence.

That this subject may be better understood, it is necessary first to consider the action of water in pipes, more particularly by a familiar use of its affections in a machine called a syphon, which is formed of two branches, as represented in fig. 1, pl. 14. The shorter branch of this instrument being immersed in any liquor, and the air drawn out, the pressure of the atmosphere on the outer surface of the fluid raises it into the tube A; when it is arrived there, its own gravity causes it to fall through the longer branch. When the syphon is employed, if the air were not drawn out of the longer branch, the equal action of the air on both ends would render it useless. It is necessary that one branch of the syphon should be longer than the other, or the pressure of the water would be equal in each tube, and would consequently support the liquor in the tube. It is on the principle of the syphon that water is conveyed to great distances by pipes laid in the ground, over obstacles of considerable height, if the intervening elevation be not more than about thirty-two feet above the level of the water at its source; because the pressure of the atmosphere will of itself support a column of water of about that height. The pipe, which immediately conducts it to its destination, must be lower than that by which it is elevated, otherwise it will not discharge itself at the given stations. Let D, fig. 1, pl. 15, represent a hill from which water is to be conveyed to a lower situation, as at B, over another hill c. The longer arm of the syphon is represented by c B, the shorter arm by c D; the water will readily flow from one station to the other, because c B is lower than A D and c, but about thirty-two feet above the level of the water at A. Its velocity will be equal to what it would acquire through a pipe placed directly from A to B, the force being in proportion to the height of the fluid at the fountain head.

Of Springs and Fountains.

We know that artificial fountains result from the laws and agents we have been considering; so likewise may natural fountains and intermitting springs be occasioned by the same affections and operations of water.

Suppose A, fig. 2, pl. 15, to represent a reservoir; B C D the pipe through which the water passes. In running from B to c, it will acquire so much velocity from its weight, as to raise it by the force gained to E, its level in the reservoir; but flowing at D, it will do so with an increased rapidity into F. This may explain the cause why some natural fountains, branching out before they have obtained the level, fall in a shower of water.

The syphon offers a solution to the phenomena of intermitting springs, naturally and satisfactorily accounting for their fluctuations.

Let fig. 3, pl. 15, represent a hill, in which is a cavity A B C D, whence proceeds a channel E F G, in the form of a syphon. The rain falling abundantly on the side of the hill, percolates through the crevices; and when the cavity is filled to the level H H, the water will be raised to F in the channel, and run through it to K, where also suppose a well or cavity. When things are thus circumstanced, the water in K may rise to the top ; but if there be no rain for a considerable time, it may sink to the bottom of it. If, instead of these irregularities, the rise of water in a well be uniform, like the flux and reflux of the sea, it may be accounted for thus: supposing the channel of communication to be as in fig. 3, when the water gets above the level of H H, it will rise rapidly, and as quickly sink down again, by means of another syphon in the receiving part or spring, and thus decrease. Fountains which depend on the solution of quantities of ice must necessarily be intermittent.

There is a spring at Engstler, in the canton of Berne, subject to a daily and annual intermittence. It begins to flow in May, but chiefly in the night.

The cause of the reappearance of the high tide in this fountain is thus explained by the learned Dr. Charles Hutton: " At this time " the mass of earth gets sufficiently warm to dissolve the great quan- " tities of ice in those parts, when various currents begin to flow " from them, though their upper surface remains unaltered." The doctor attributes the diurnal changes in the mass of earth covered by the ice to the presence and absence of the sun. " This portion of " the earth must imbibe the sun's rays for a very considerable time " before it can have sufficient heat to dissolve the ice. Hence it is " that the liquefaction of the ice does not take place till after sun- " set; and in some parts it does not appear till near twelve o'clock at

" night, on account of the distance it came from," &c. " No doubt,
" all the varieties observed of intermitting wells and fountains
" depend on the principle of the syphon combined with the sudden
" fall of rain, or liquefaction of snow and ice," &c. Before the use
of the pendulum was determined, time was ascertained by different
modes, but principally by the fall of water.

To enable the reader to understand how equal quantities of
water may be discharged from a vessel in equal divisions of time, 1
shall exhibit a curious clepsydra described by the doctor in his
" Philosophical Recreations," and invented by Hero of Alexandria.
Let fig. 5, pl. 15, represent a glass cylinder, on which twenty-four
equal parts are graduated; fill it with water, and place the shorter
branch of the syphon in a piece of cork, to keep it upright on the
surface of the fluid. It is evident that the efflux of water will be
uniform, as it depends solely on the unequal pressure of the air on
she surfaces of the fluid, within the two branches of the syphon.
Therefore, while it continues to flow, it does so with an equal
motion.

It is supposed by some persons that the ancients had but a slight
knowledge of hydraulics: however, from Pliny's description of his
fountains, and other circumstances related in his letters, we are
warranted in inferring that in his time the natives of Tuscany at
least were acquainted with this science, as well as with that of
mechanics.

The variety of machines used for the purpose of raising water,
precludes a general description in this lecture. Suffice it then to
say, that the action of all of these depends on the pressure of the
air, on the gravitation of water, on the mechanical force of the

engine employed, and on attraction and cohesion; of each of which, excepting the last mentioned, I have treated sufficiently in the preceding lectures for the purpose of general application. But much remains to be said on the subject of capillary attraction: this property arises from the wonderful principle of matter called cohesion. The attraction of cohesion is so powerful, that if two perfectly smooth surfaces be placed close together, great force is required to separate them. Two polished brass plates, oiled to fill up any interstices on their surfaces, will cohere so strongly together, that they cannot be parted without violence.

Cohesion takes place in vacuo, as well as in the air: for if the two plates be suspended within the receiver of an air-pump, the same force will be necessary to separate them, as in the open air. Hence we infer, that it is their natural and mutual attraction, and nothing else, which keeps them together.

Capillary attraction is most evident in tubes, the channel of which is of about a hair's breadth, though it may be seen in such as are much wider: but according to the capacity of the tube, reciprocally will be its power of raising water above its natural level; for the phenomenon arises from the near approach of the interior surfaces.

A very trifling intervening substance will destroy the effects of cohesion; as is evident on greasing the inside of the tube, when the water does not rise in it. That it is attraction which causes water to ascend in a capillary tube, may be seen by letting a drop run down its outer surface, which arriving at the bottom, enters and rises up within the tube.

The elevation and suspension of water in a fine tube are supposed by philosophers to depend on the periphery of the concave surface of the tube, to which the upper part of the fluid adheres.

The effects of capillary attraction, by which moisture is conveyed through the interstices of wood, &c. are employed by mankind in various mechanical operations, particularly in dividing substances with greater ease and safety than by percussion. A very striking use of this property presents itself to my recollection, as employed in dividing mill-stones. The stones used for this purpose are first formed into cylinders of considerable length; to separate these into the proper proportions, indentures are cut at suitable distances on the outer surface of them in a circular form, into which are driven wedges of dry wood. By the application of water to these wedges, the cylinders are divided; for the fluid is drawn to the other extremity of the wood by the capillary attraction, which causes an expansion that splits the stone asunder. Capillary attraction is concerned in most of the operations of nature; for nutriment is conveyed through these fine tubes to all the parts of vegetable substances and animal bodies. Many familiar effects arise from corpuscular attraction; as, oil supplying the wick of a lamp, water dividing sugar or salt: in a word, all the pores and interstices of bodies are capillary tubes, which imbibe fluids, effluvia, &c.

Let us employ our knowledge of hydraulics by applying it to understand its most perfect uses; and on this subject, as on all others, glorify our Maker, by contemplating his exquisite hydraulic contrivances in the system of the highest order of animal nature. In this sublime contemplation that grand hydraulic engine the heart, which works the whole machine, first presents itself; whence, through pipes of various sizes, the animating blood is conveyed to their extremest verge; and, by gentle meanderings, returns through finer pipes to its grand repository. No one can doubt that the circulation of the blood is necessary to animal existence; yet we do not always reflect on the supreme skill and contrivance exhibited in this curious process; which, though complicated, is simple; for the direction of the pipes that conduct the blood to the different parts of the body, and of those which convey the returning current to the heart, may be perfectly understood, and their tendencies comprehended even by human reason. In the hydraulic engines invented and employed by man, it is sufficient that they are active in their emissions: but this machine of God has to perform two opposite and contrary offices; it not only discharges in all directions, and in due proportions, but it receives all it has discharged, for the purpose of continuing the action or circulation. For which wise and indispensable purpose, two distinct systems of pipes are laid in the animal body, arteries and veins —by the former the blood is propelled with vigour from the heart to all parts of the body; and by the latter its returning current is conveyed back to its original source.

The construction of these two systems of pipes is equally expressive of the divine Mind with every other part of this curious mechanism. The blood, in its diffusion from the centre, passes through large tubes, called arteries; from which it is propelled into others gradually less and less, till it arrives at their extremities; nourishing and refreshing the various parts of the body in its progress; then, uniting with the veins, it returns, by the force acquired in its projection, to the heart.

As the vital fluid is thrown into the arteries with considerable impetus, those vessels are formed and constituted to sustain much greater force than the veins; which latter receive the blood but in small quantities, beginning in fine, and terminating in large, channels. Hence the Creator has made the former not only much stronger than the latter, but has guarded them with fences within the body; while the veins lie nearer to its surface and are less protected, not being liable to the same injury as the arteries. So remarkable is this care of the Divinity, that grooves are formed in bones to receive the arteries, and sometimes entire hollows left in them for the free passage of these vessels.

The action of the heart, which causes the circulation of the blood, produces its effect by alternate or reciprocal contraction and relaxation. In the centre of this organ, which is placed in the middle of the body, lies a hollow muscle, invested with spiral tubes intersecting each other in both directions. Into these tubes are inserted the great arteries that convey the blood out; and the veins which receive its returning current. By each contraction of this muscle the blood is forced into the arteries, and by every expansion an equal portion is returned to this receiver. Hence at every beat of the heart an exchange of blood takes place in a certain degree, the quantity being nearly an ounce in weight, which is what the muscle holds. The weight of blood in a middle-sized person is supposed to be about twenty-five pounds; and this quantity passes through the heart once every four minutes. The heart contracts and dilates about four thousand times in an hour, and thus there passes through it about three hundred and fifty pounds of blood in that time.

Though it is partly by the contraction and dilatation of the muscle above mentioned that the blood is injected and received by the heart, yet these are not the only means by which a due balance of circulation is preserved; for the lungs also assist in regulating this indispensable process. Air being necessary to the constitution of the blood, to enable the heart to form a communication with that element, it is situated in the bottom of the lungs, which are supplied with numerous air vessels; these lie close together: after the blood is poured into the heart by the veins, before it is again propelled through the arteries, it is carried by an artery to the lungs, filling their vessels; then, after receiving the impressions of the air, it returns to its source by a large vein, and is thence forced with a given impetus, by the action of the heart, into the system, to impart its salutary invigorating influences.

The heart has four conducting pipes;—two called ventricles, which send out the blood, one into the system, and the other into the lungs; and two auricles, which receive the blood from the veins and the lungs. The receiving cavities communicate with the discharging ones, into which, by contraction, they force the blood; whence it is propelled, by the discharging cavities, into the arteries. The regular effects and dilatations of the heart and lungs must depend also on the action of valves, which, opening in one direction only, prevent the fluid from returning by the way it enters. The same action of the heart that carries the flow of blood into the mouth of the artery, were it not for valves, would stop the return of the current into the heart from the veins: the dilatations of the parts would also cause the blood to flow back whence it was projected. But the valves, one of which is placed between each auricle and its ventricle, and another at the mouth of each great artery, effectually prevent this mischief. These valves are composed of fine ligaments; yet they endure for the longest life of man without injury, and without impediment; for they are the work of God, continually supplied, by his wisdom and goodness, with strength and power. The heart is inclosed in a membranous bag, that guards its surface, without confining its action. In this bag is a small quantity of water, sufficient to keep the surface of the heart flexible and moist, which no doubt is necessary to its constitution and functions.

Our blood is supplied by the aliment which nourishes our bodies; for a portion of our food, when converted into chyle, mixes and assimilates with the blood already formed; which operation is equally wonderful with those we have previously contemplated, yet so enveloped in combinations, that it requires ocular demonstration of the parts of the body employed in these operations, perfectly to comprehend it.

I shall endeavour to explain one operation more in the system, as its simplicity will admit of a perfectly satisfactory theoretical description. It is the admirable contrivance of the pipes that serve the purpose of swallowing and respiration, which, though not properly hydraulic instruments, are too important to escape our notice, when treating of the vessels of the body immediately connected with it.

The larynx is composed of two pipes, both opening into the mouth; through one of them the food passes to the stomach, and the other, which leads to the lungs, is for the purposes of breathing and speech. To prevent any inconvenience arising to either from the operation of the other, at the top of the wind-pipe a valve is fixed, which shuts when the food passes, and immediately after opens for the purpose of respiration. We may perceive the necessity there was for this contrivance, by observing the trouble and distress that sometimes, though seldom, happen from swallowing too quickly; for if but a single drop or atom get into the wind-pipe, a convulsed motion is excited, which continues till the offending cause is removed. The extreme irritability of this part is beneficial in warding off danger.

Here let us stop; for endless is the theme—to enumerate the exquisite contrivances, the exalted wisdom and kind beneficence exhibited by each operation in the wonderful system of the human frame!

PLATE XVI.

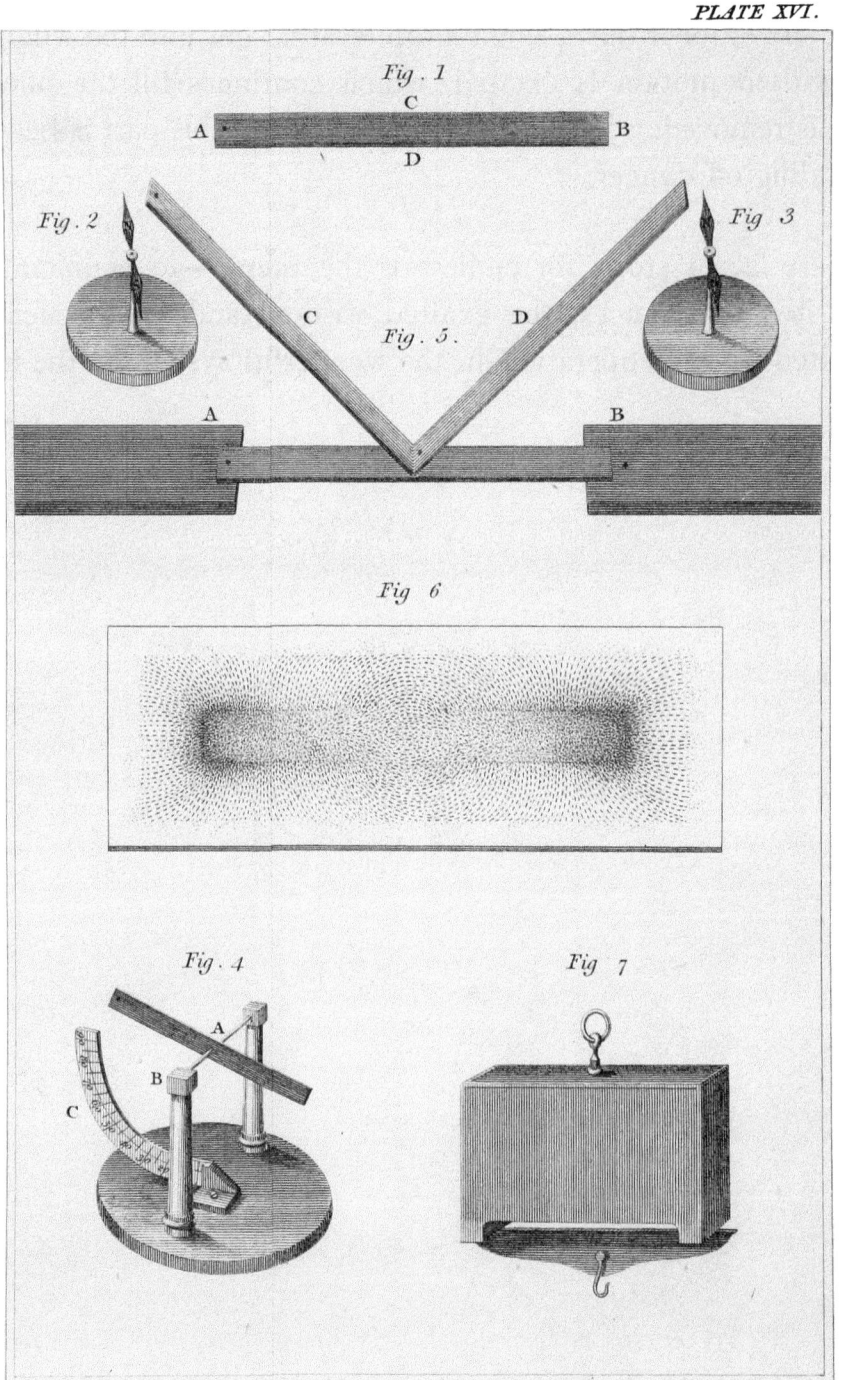

Fig. 1

C

A   B

D

Fig. 2   Fig. 3

C   Fig. 5.   D

A   B

Fig 6.

Fig. 4   Fig 7

A

B

C

S. M. Bryan design.   H. Mutlow sculp.

PLATE XVII.

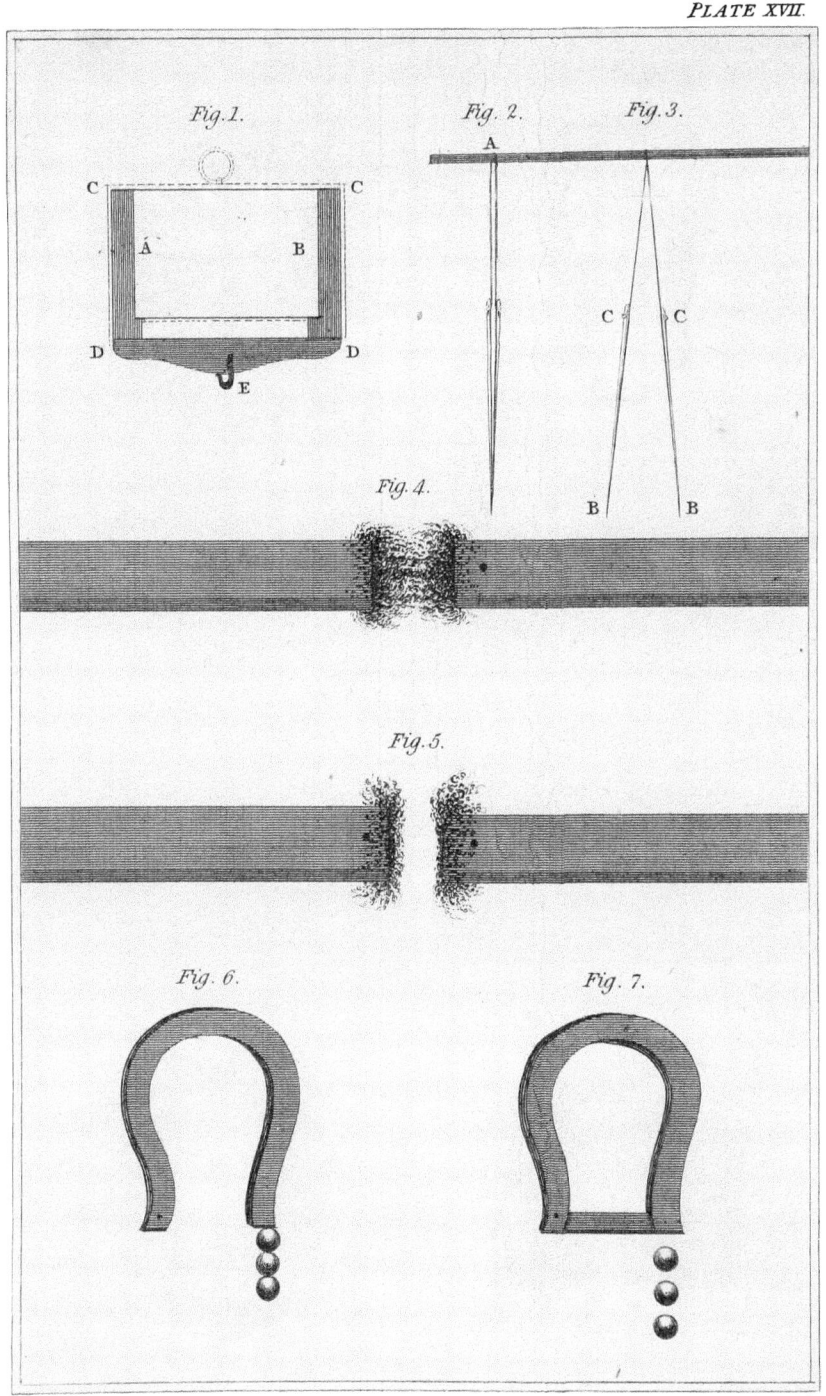

Fig. 1.

Fig. 2.

Fig. 3.

Fig. 4.

Fig. 5.

Fig. 6.

Fig. 7.

S.M.Bryan design.

H.Mutlow sculpt.

PLATE XVIII.

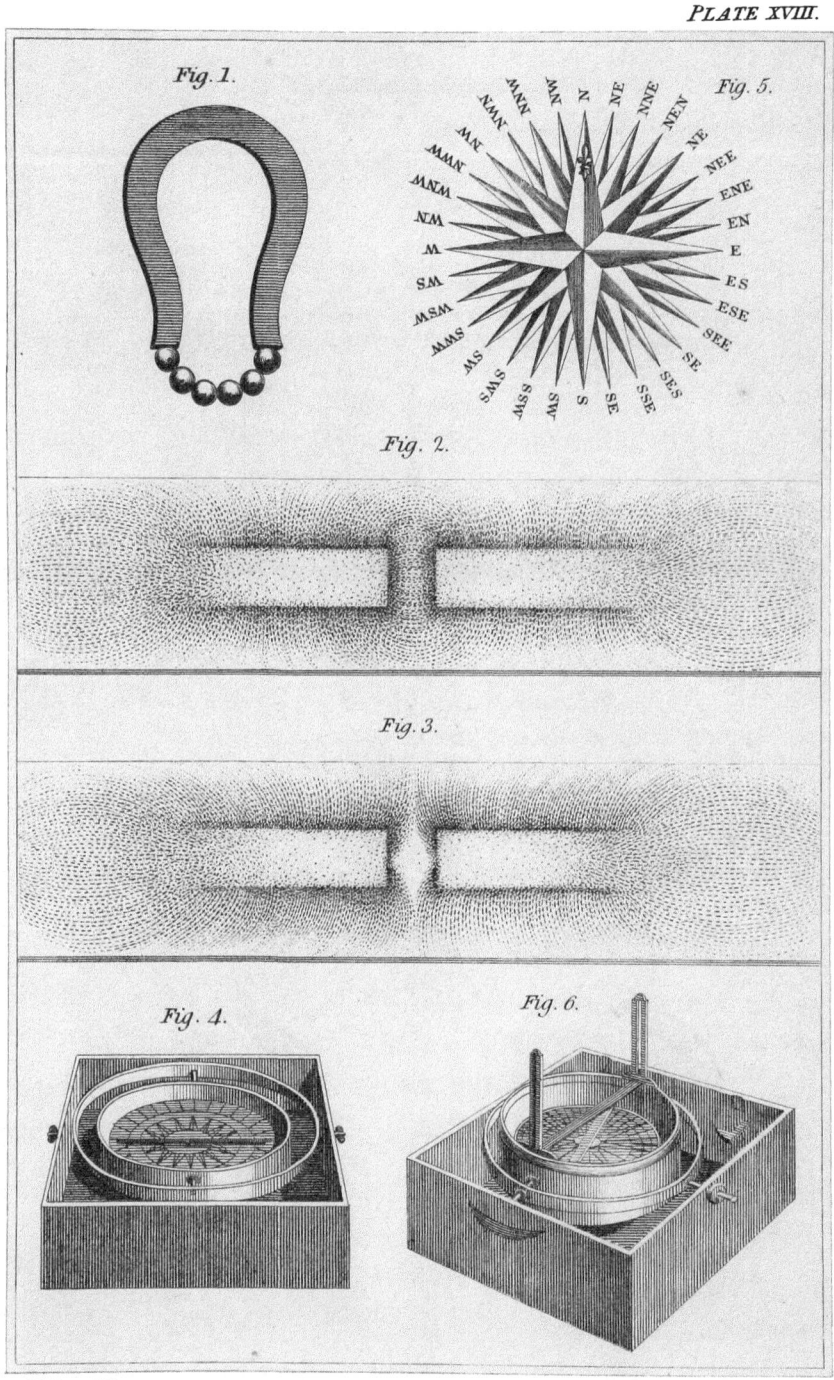

Fig. 1.

Fig. 5.

Fig. 2.

Fig. 3.

Fig. 4.

Fig. 6.

S. M. Bryan delin.

H. Mutlow sculp.

LECTURE THE SEVENTH.

ON MAGNETISM.

The magnet's potent spell attracts the ore,
Whose strong affinity obeys its power;
Possess'd, diffus'd, its laws impress'd exacts;
The needle points where'er its power directs;
That power, so useful to commercial store,
Points the skill'd seaman to the distant shore:
Thus, may the force of virtue charm my soul,
Possess my reason, and my act controul:
Nor useless be the boon—but teach me this,
To point the mind of youth the way to bliss;
Instruct me how to teach life's vernal dawn,
By strong impressive science, truths unknown,
Where Wisdom, Virtue, Truth unrivall'd shine;
For nature's laws result from pow'r divine;
'Till pure, exalted, their chief joy they find,
In adoration, praise, and love combin'd,
To God Omnipotent! whose gracious word,
Created all!—and saw that all was good!

AMONG the various powers applied by our wise Creator to effect his purposes, none excite more astonishment than those of Magnetism; which like all the others are known only by their effects— effects useful, peculiar, wonderful!

The natural magnet is a solid mineral substance, of a dark greyish colour, and of a compact and weighty nature. It is found in different soils and situations, but chiefly in iron mines, and possesses the powers of attraction and direction. The artificial magnet is a piece of iron or steel, to which the properties of the natural magnet have been communicated. The name magnet is supposed to have been derived from Magnesia, the province in which the effects of the loadstone were discovered. A true magnet, whether natural or artificial, has the following characteristics :—it attracts iron, and points nearly to the poles of the world ; possesses both an attractive and repulsive power within itself; and always inclines or tends to a point below the horizon.

The ancients were totally unacquainted with the nautical use now made of the magnet, having only discovered one of its properties, that called attraction. To Columbus we are indebted for a great part of its present extensive usefulness in navigation ; for which benefit his memory must be revered by all lovers of science, and particularly by those persons who are benefited by commercial advantages. The essential properties which cause the phenomena of the magnet have not been ascertained ; yet those conjectures formed on the subject, which ascribe its properties and affections to a subtile effluvia, universally disseminated through the earth and its atmosphere, and produced from a central body of a spherical form, appear to be well founded in reason, and are also confirmed by experiment. But the cause of its directive power, and the variableness of its direction, appear to be almost inscrutable.

The magnet has no particular form, or distinguishing external marks, but appears like a stone. Meteorologists have extracted iron from it, but in such scanty proportion, as not to pay the expence of fusion. Modern chemistry has discovered that iron, in its oxyde state, pervades all nature ; but the magnet attracts it only in its metallic form.

We will now proceed to examine the known properties and laws of magnetism; in which useful science we shall find much in the extreme subtilty of its nature to admire, much in its elaborate affections to amuse, and in its results every thing to excite our admiration, astonishment and gratitude.

We are already acquainted, by our former investigations, with five kinds of attraction :—First, gravitation, which enables all bodies on the surface of the earth to retain their situations ; and, combined with the centrifugal force, causes all the planetary bodies of our system to revolve round the sun at certain distances from that luminary and from each other: secondly, cohesive attraction, which keeps the parts of bodies together, and unites them in close compact: thirdly, chemical attraction, called affinity, which causes certain bodies to distinguish each other in preference to other substances introduced into a compounded mass, and to unite together: fourthly, capillary attraction, which causes fluids to rise in very small tubes (this may be connected with cohesive attraction, being only a different effect perceived of the same cause): fifthly, we have magnetical attraction; the affections of which the experiments I shall have the pleasure of exhibiting will explain.

Let fig. 1, pl. 16, represent a piece of steel, to which a magnetic property has been communicated. The two extremities, A, B, are called the poles of the magnet; the line from c to d, its equator. Suppose figures 2, 3, to be magnetic needles, freely suspended, pointing north and south. Fig. 4, is a dipping needle; A is the

U

axis of the magnet on which it freely turns and dips below the horizontal line B.   c is a quadrant, to shew its depression or inclination.

The tendency of the needle to the north and south, is called its direction.   Its variation from due north and south, is called its declination; and its dip below the horizon, its inclination.

Experiment of communicating the Magnetic Virtue.

The magnetic virtue may be communicated to a bar of iron or steel, by placing two natural magnets, A, B, fig. 5, in a straight line, the north end of the one opposite to the south end of the other; and at such a distance, that the two ends of the bar to be touched may rest separately upon them: the end designed to point north resting on the south pole of the bar, and *vice versâ*.   Two other steel bars must be placed in such a manner, that the north end of one and the south end of the other may rest on the middle of the horizontal bar, the end of each being elevated so as to form an acute angle with it, as C, D, fig. 5.   The two oblique bars should be separated, by drawing them contrary ways along the cross bar, towards the natural magnets, keeping them at the same elevation all the way; when, removing them from the cross-bar, and bringing their north and south ends in contact, then applying them again to it as before, and repeating this four or five times; after which, performing the same operation with the other surface of the cross-bar, it will have acquired a permanent magnetism and polarity   Small needles for compass boxes do not require this process, but may be rendered magnetic by friction, merely passing them three or four times over a magnet in one direction.

A compass needle while receiving the magnetic virtue is violently agitated; but when it has fully acquired the property, the agitation ceases. A magnet loses nothing of its own strength by a communication of its property to other bodies, but gains some addition to its power by the performance. A north or south pole of a magnet, when applied to a bar or needle, produces the contrary polarity; therefore two magnetic bars should not have the poles of the same description placed together, for that position will diminish their individual power.

Each point of a magnet may be considered as the pole of a smaller one, tending to produce on the points of the magnet a force contrary to its own. The degree of this effort will be greater in proportion to the force of the point, and its nearness to the poles on which it acts; hence, a narrow and long bar of steel is more powerful than a short and broad one.

Whatever may be in reality the cause which produces magnetism, we see that its nature is very subtile and active; by its passing through substances of the most compact nature, and by its virtue remaining unaltered.

Experiment on the Circulation of the Magnetic Fluid.

Placing a magnet under a plate of glass, covered with white paper to show the experiment, and then strewing it with steel filings, the particles of steel will arrange themselves in a form, represented fig. 6, pl. 16; which points out the course of the circulating element, and their position determines the magnet's greatest force to be in the poles, because over them the filings rise perpendicularly. This increase of power at the poles may be in consequence of the points being further removed from the influence of the two powers on each other, which must necessarily diminish their individual strength.

This experiment seems to confirm the idea of a subtile fluid flowing from one pole to the other, and circulating round the magnet.

A piece of iron or steel brought near to a magnet, is attracted by it; and when in immediate contact they adhere together with considerable force. It is not the magnet alone which attracts, for the magnet and iron mutually attract each other: when either is held and the other left at liberty, the latter approaches the former.

Experiment on Magnetic Attraction.

This fact may be proved by placing a magnet on one piece of cork, and a piece of steel on another, and floating them on water; when, both being unconfined, they will approach each other: and on holding the piece of steel in the hand, the magnet will approach to it with the same velocity as they approached to each other when both were at liberty.

It appears from the foregoing experiment, that the iron by being placed near the pole of a magnet becomes possessed of a contrary power. Their mutual attraction may also be explained by the laws of action and reaction, which are always equal and opposite to each other.

Neither magnetic attraction nor repulsion is affected by an intervening body; but heat weakens the power of magnetism, and sometimes destroys it: yet its property may be restored, though not its power in the same degree as before. May not this circumstance arise from some of the effluvia having gone off in consequence of heat? Iron when red hot is not attracted by the magnet; perhaps its whole affinity with that power has evaporated.

Philosophers have in vain endeavoured to estimate the force with which the magnetic attraction acts at different distances; but as that law has not yet been fully ascertained, all that we can infer from their observations and experiments is—that the magnetic power extends further at one time than at another, and therefore its sphere of action is variable.

A magnet cannot support even its own weight of metal, but its power may be much increased by means of arming, which is thus performed:

To arm a Magnet.

Cut the magnet into a regular parallelopipedon (like A B, fig. 1, pl. 17), and let its two poles be parallel planes: place this magnet in an armour of soft iron, as represented by C C, D D; which, having a cross-piece, E, with a hook attached, will support great weights suspended from it. The advantage gained by arming is very considerable; a magnet that will of itself support four or five ounces, will when armed sustain twenty times that weight. A section of the armed magnet is exhibited, fig. 1, pl. 17: C C, D D, represent two plates of iron, of a thickness proportioned to the size of the magnet. A magnet and its armour may be enclosed in any material excepting iron. Fig. 7, pl. 16, represents the magnet armed, and enclosed in a case.

The power of a magnet may also be augmented without arming.

by simply introducing another piece of iron below that it at first
supports; as is evident on presenting to it a piece of iron heavier
than it can sustain, and afterwards holding under it another piece
at a small distance from the former, when the magnet will support
what before it could not lift.   The cause of this is assigned by Ca-
vallo to the last piece becoming magnetic, and so increasing the
attraction of the first piece, and in the following manner.—The end
of a piece of iron which is presented near the north pole of a magnet
becomes possessed of the south, while the other extremity possesses
the north polarity.   Again; the second piece being held near to the
north pole of the first piece of iron, acquires a south polarity.   This
must increase the north power of the first piece, when its south
power must also be augmented in the same degree, and thus it is
that the magnet supports a greater weight by the communication.
That this is the true cause of its increased power of attraction is
evident by placing the south pole of another magnet below the
piece of iron; when the same effect takes place.   Presenting the
north pole of a magnet to the first piece of iron produces a contrary
effect; for it diminishes the power of the first magnet.

Experiment on Repulsion.

Let fig. 2, pl. 17, represent two large needles, suspended from
A by threads.   Present the north pole of a magnet just under
them, and they will repel each other, as exhibited fig. 3.   This
phenomenon arises from the known effects of magnetism ;—namely,
the two ends, B, B, of the needles become possessed of the contrary
power to the bar presented, or with south polarity, but each with
the same power; hence they repel their own property: the north
poles or upper ends c c, being also of the same power with each
other, exhibit the same repulsion.

Experiment on the increasing Power of a Magnet.

Suspend a magnet by a hook from some fixed point, and attach as much iron to it as it will support together, with a scale, which must also be affixed; and you will find that every day you may put additional weight in the scale, and the magnet will support it; which shews that its power is constantly increasing.

It is supposed that the iron, becoming magnetic, increases the power of the magnet in the manner before described. When the iron is removed from the magnet, the power of the latter is rendered weaker than it was before the experiment was made. This illustrates the theory of Æpinus, that the magnetic fluid is unequally distributed in a magnet which has a fixed polarity, one pole being overcharged, while the other is undercharged with it; and that there is always a strong attraction between these contrary poles, in consequence of this unequal distribution of power: but when a piece of iron is presented to either, that, by its becoming possessed of a contrary polarity to that of the magnet, the power of each end on the other is weakened by the communication, and thereby its individual power increased; for there is in every magnet a strong attraction between its poles: but when another substance, or a magnet, is presented to either, the effect is stronger by being drawn from the contrary pole. Hence we may suppose that a magnet becomes continually weaker when left alone, so that it is necessary either to place it in armour, or leave a piece of steel or iron on its poles; because at these points the powers are at the greatest distance from each other's effect.

It is not more extraordinary than true, that the magnetic power may be acquired without the application of a magnet, and by friction be made to communicate that power to iron or steel. Rubbing one piece of iron on another will produce evidences of the magnetic virtue; and even a certain position of either, long continued, will render that effect permanent. The famous philosopher of our country Dr. Gilbert, in the sixteenth century, observed that the small bars of a window which were placed obliquely to the horizon, and nearly north and south, by remaining in that situation for many years became magnetic. The polarity thus communicated may be from the earth and its atmosphere; for all the effects of magnetism evince that the power is derived from those sources, though the peculiar directive power cannot be traced to its primary natural cause. The particles of iron being universally diffused—through all animated nature, as well as in all substances in the earth—may not a magnet have some effect on the animal economy? As this universal diffusion of iron fully justifies the idea that the magnetic fluid is one of the elements of the earth and its atmosphere, may we not also conceive the magnetic effluvia to be equally disseminated through the globe, in such bodies as do not exhibit any evidences of its existence; and that its visible effects result from that equilibrium being destroyed?

Experiment on the different Power of Attraction in different Parts of a Magnet.

Place a dipping needle on a magnet, and move it from one pole to the other: when it is over either of the poles, it will stand perpendicularly to the axis of the magnet; when exactly between the two poles, over its equator, the needle will be exactly parallel to the bar. This shews that at the poles the individual attraction is strongest; the fluid being in equilibrium at the equator of the magnet, neither attraction nor repulsion can individually act at that part. Between the equator and the poles the needle inclines to the bar in different angles according to its distance from either of them.

Experiments on the Action of the Poles on each other.

The dipping-needle serves to shew the action of the two different poles on each other; for on presenting the north pole of a magnet to the south pole of the needle, it is attracted; but if we present the same pole of the magnet to the north pole of the needle, it is then repelled and flies from the magnet. Strew steel filings on a pane of glass, and put the north pole of a magnet under it, they will then rise on the paper; but on holding the north pole of another magnet directly over these filings, they will immediately fall. Dip the north pole of one magnet and the south pole of another in steel filings, and bring the ends of the bars toward each other; then the filings will unite, as exhibited fig. 4, pl. 17. But dip the two north poles and bring them in contact, and the filings will recede from each other, as represented fig. 5, pl. 17

Two magnets placed in a straight line at a small distance from each other, the south pole of one opposed to the north pole of the other, with a pane of glass over them; on sprinkling steel filings, and tapping the glass to produce a little motion in the filings, they will arrange themselves in the direction of the magnetic fluid; those lying between the two poles, and near the axis, being disposed in straight lines, going from the north pole of one magnet to the south pole of the other; as seen at fig. 2, pl. 18. Reverse the order of the magnets, by placing the two poles of the same name opposite, when the filings will be arranged in curves receding from each other, in the manner represented by fig. 3, pl. 18.

Experiments on the Magnetic Effluvium, endeavouring to place itself in equilibrium by the shortest way.

The effort the magnetic effluvia appear to make to attain an equilibrium by the nearest way, is, I think, made evident by the following experiments. Take a horse-shoe magnet, like fig. 6, pl. 17, and with its north pole raise three or more balls of soft steel; when they will remain suspended from it, as at fig. 6, pl. 17, because of the power exerted by the magnetic fluid to diffuse itself: but touching first the south pole with a small piece of steel, and then bringing it in contact with the other pole, the balls will drop, as seen at A B, fig. 7, pl. 17, because the fluid can form its circuit a shorter way.— For the same reason, when each end of the magnet has supported two balls, on placing two more below these, they will all unite in a semicircle, as exhibited fig. 1, pl. 18; that being the nearest line of communication they can then form.

Of the Declination of the Needle.

The north pole of the magnet, in every part of the world, points nearly north; yet it very seldom shews that direction exactly. Hence the magnetic meridian seldom coincides with the observed meridian of any place on our globe, but generally varies either to the east or west. This variation is not uniform at different places, nor does it always agree even in the same place: at London, for instance, in the year 1640 it was 11° east, but it is now about 23° west. This variation is always reckoned from the north, either east or west. The directive power of magnetism, though generally exhibited by a touched needle, is also evident in very small bars of steel or iron freely suspended; as may be seen by fine pieces of either floating on the surface of water: but to exhibit this property, they must remain some hours, when they will point nearly, if not exactly, north and south.

The directive property of the magnet, according to Dr. Halley's hypothesis, is supposed to arise from the current of the magnetic fluid issuing from a central magnetic globe, which, passing through the earth and its atmosphere, causes light bodies to move with it.

To account for the direction of the magnet being variable, and this variation not regular at the same place, nor in an uniform degree at the same time at different places, various hypotheses have been formed, and some truly curious and interesting experiments have been made to illustrate them, of which number the following appears the most ingenious and satisfactory.

Messrs. de la Hire, senior and junior, formed a globe out of a very large magnet, and, by suspending it, found its poles; they next traced out its equatorial and meridional circles. The globe was about a foot in diameter, and weighed one hundred pounds. Placing it due north and south, and in a position that answered for the latitude of the place of observation, they perceived its declination east and west, in regard to the situations of places on it. From these remarks they inferred that the magnetic fluid is diffused through the whole earth, and obeys the universal laws of magnetism; yet they do not explain the causes of the different variations of it at the same place. The regular declination observed on the magnetic globe was owing to the equality of contexture in its parts, and the varying magnetic force at different places on its surface. But as the contexture of the earth is very irregular, perhaps that circumstance, united with the numerous processes carrying on within it, is the cause of the variation. Perceiving that the regular variation on the magnetic globe arose from its uniform contexture, we may infer that the inconstancy of the variation of the needle on the globe of our earth arises from the inequality of its parts. No perfectly satisfactory hypothesis having yet been formed respecting the variation of the needle that can be authenticated by facts, it is impossible to foretel what this irregularity will be at a future time at any particular place, or other circumstances depending on that knowledge, though derived from the experience of a long-continued series of observations.

The ingenious Mr. Canton discovered a new variation of the magnetic needle, which he communicated to the Royal Society. Observing the direction of a touched needle for a whole day, he perceived that it was never perfectly at rest; that its western declination from the pole was greatest in the morning, and least at night; and at about noon in a medium of its diurnal variation. He offers the following rational solution of these phenomena, founded on the known fact—that a magnet when heated loses something of its natural force. He supposes the direction of the needle to be occasioned by the attraction of the magnetic fluid, and that the attraction is strongest where the heat is weakest; therefore, that the needle at sun-rise with us is not so forcibly impelled towards the east, because the magnetic force is lessened by the sun's influence; consequently the needle points rather more westerly at that time. When the sun is on our meridian, the variation is not changed, the action of the sun on each side of us being then equal: towards evening the needle points more easterly, because it naturally points to the part within its range the least heated by the sun.

Experiment.

This effect may be understood by heating a magnet, and placing it on one side of a needle, and another magnet in its natural state on the other side, when the needle will decline from the heated one.

Mr. Canton perceived, from repeated experiments, that the diurnal variation of the needle was about 20 minutes of a degree, from sunrise to sun-set.

Of the Dip of the Magnetic Needle.

The needle has a dip, or inclination; the cause of which, like every other peculiar characteristic of this curious phenomenon, is unknown. It may be seen, by placing an untouched needle on a pivot, and presenting a magnet to it, when it will incline towards a point below our horizon. To counteract this effect, the mechanist who constructs compasses, files off part of the inclining end, and by that means balances the needle on the pivot. The inclination of the needle is as variable as its declination. It also varies at different parts of the earth at the same time. The idea of the inclination having reference to latitude only is a mistake, it being as irregular in that respect as the declination; for at Paris in 1800 it was $72°$ $25'$ north, and at Siena $18°$ south. No doubt, these variations depend on the same causes as those of the direction of the needle.

Theory of Magnetism.

The whole that can be inferred of the nature of the phenomena of the magnet, is briefly this;—that it attracts bodies in the earth; and that it has a directive power which is variable, arising perhaps from the unequal diffusion of the magnetic power in the earth and atmosphere, depending on the different constitutional circumstances of each of them, together with the effects of heat and cold on that power. Its attraction is evident on bodies on the earth; and we know that the earth contains bodies of this attractive nature, for from the earth they are procured; and we must suppose its direction depends on the inequality of attraction in the earth. The variation in that direction may also depend on the parts which contain the attractive power being more or less heated. These natural and hidden causes being incalculable by us, we never must expect to arrive at a perfect knowledge or estimation of them. The magnetic fluid may be either formed of two kinds of elements united by affinity; these elements having a greater tendency to each other than to themselves: or the phenomena perceived of attraction and repulsion, in the former case, may be produced by the endeavour of the disturbed effluvium to place itself in equilibrium, and in the latter from its natural repulsion to itself. The directive power of the needle, and the mode of constructing compasses, are so well known, that it would be superfluous to introduce them here. Fig. 4, pl. 18, exhibits the mariner's compass; fig. 5, the compass card; and fig. 6, the azimuth: which latter compass differs from the other only in having two sights, instead of one, to observe the sun and its azimuth, in order to ascertain the variation of the needle at the time and place by that means.

Let us conclude our observations on the magnet by religious and moral inferences. We find the most evident effects of Infinite Wisdom cannot be traced to their first principles by finite reason: why then should we attempt to understand the nature of spiritual existences or discredit the truths of revelation, the sublimity of which must and should be beyond our conception and comprehension? Purposely has God ordained them so, to exercise our faith, and excite our attention to the other duties of religion. Let us then regard these holy mysteries as evidences of the love of God to his creatures, as well as heralds of his pure and perfect intelligence. Let the attractive graces of the Gospel impel us towards the goal of happiness; for unless our inherent excellence be corroded by the rust of scepticism, or evil deeds, the precepts of our holy religion cannot fail to have due influence on our nature, which they are formed to attract by their mild dictates, to direct by their genial influence, and to govern by their steady, harmonious and undeviating laws—laws at once impressive, benevolent and just!

PLATE XIX.

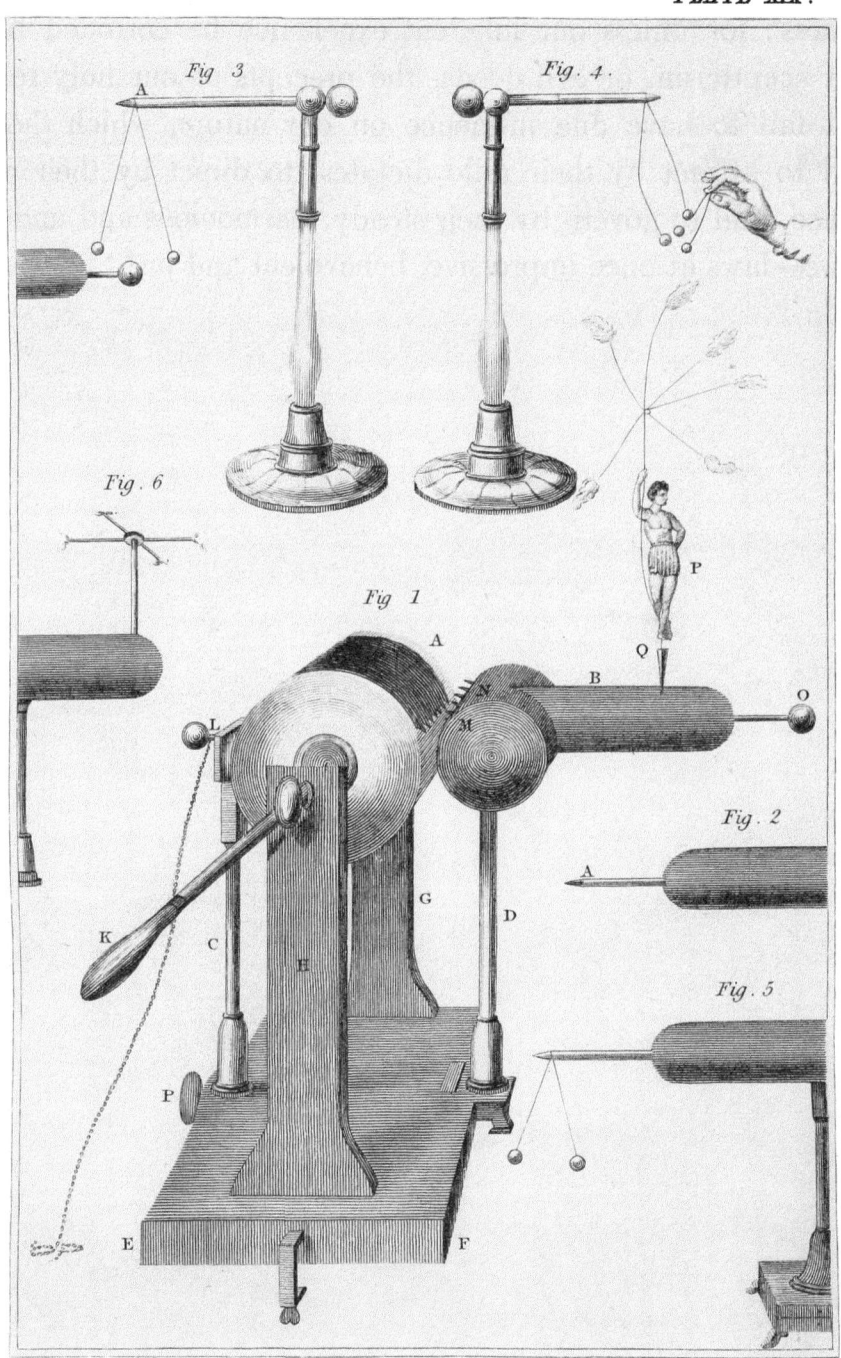

Fig 3

Fig. 4

Fig. 6

Fig 1

A

B

N

M

L

O

K

C

G

D

H

Fig. 2

A

P

Q

P

Fig. 5

E

F

T. Noble delin.

H. Mutlow sculp.

PLATE XX.

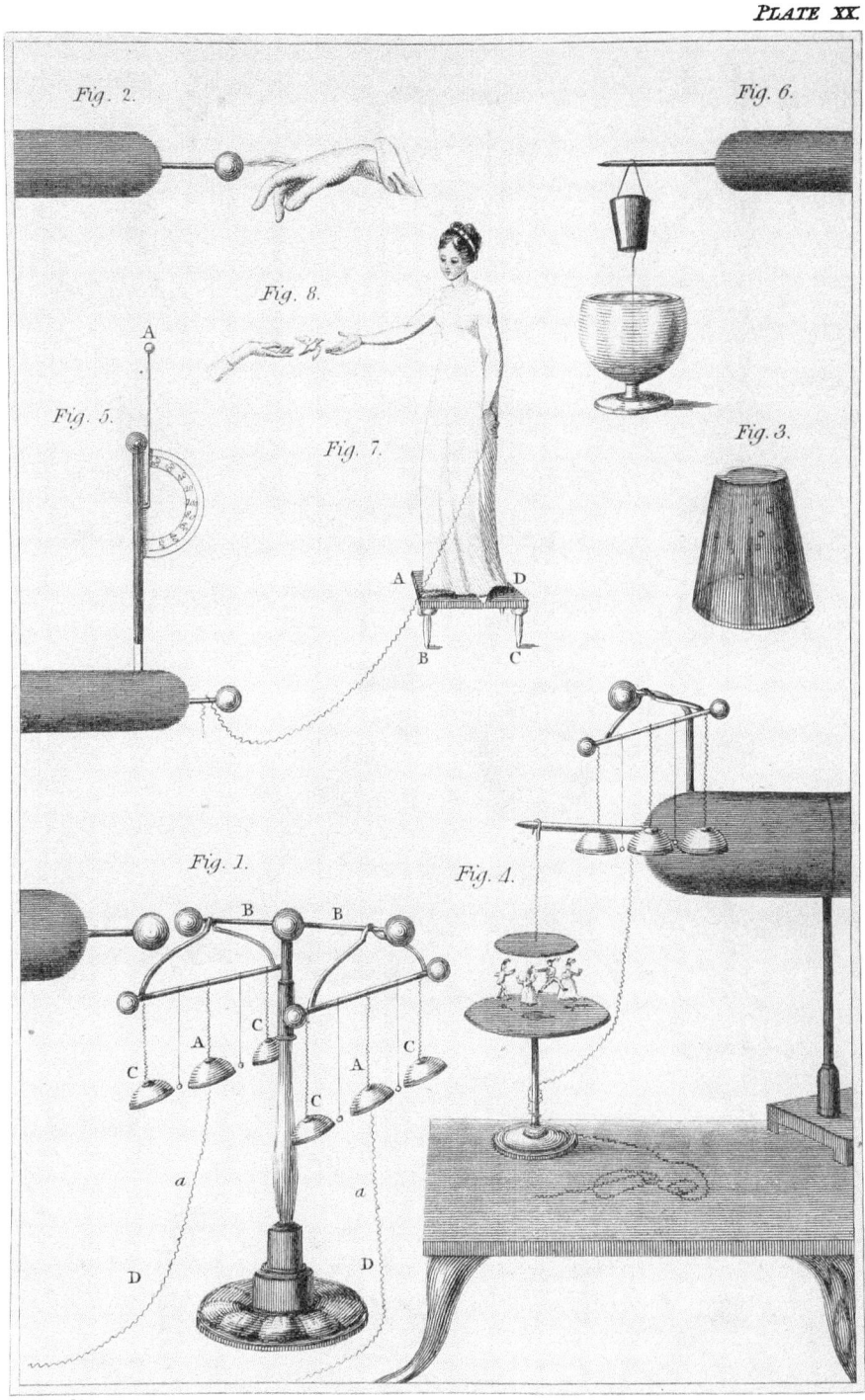

Fig. 2.

Fig. 6.

Fig. 8.

Fig. 5.

Fig. 7.

Fig. 3.

Fig. 1.

Fig. 4.

T.Noble delin.

H.Mutlow sculp.

PLATE XXI.

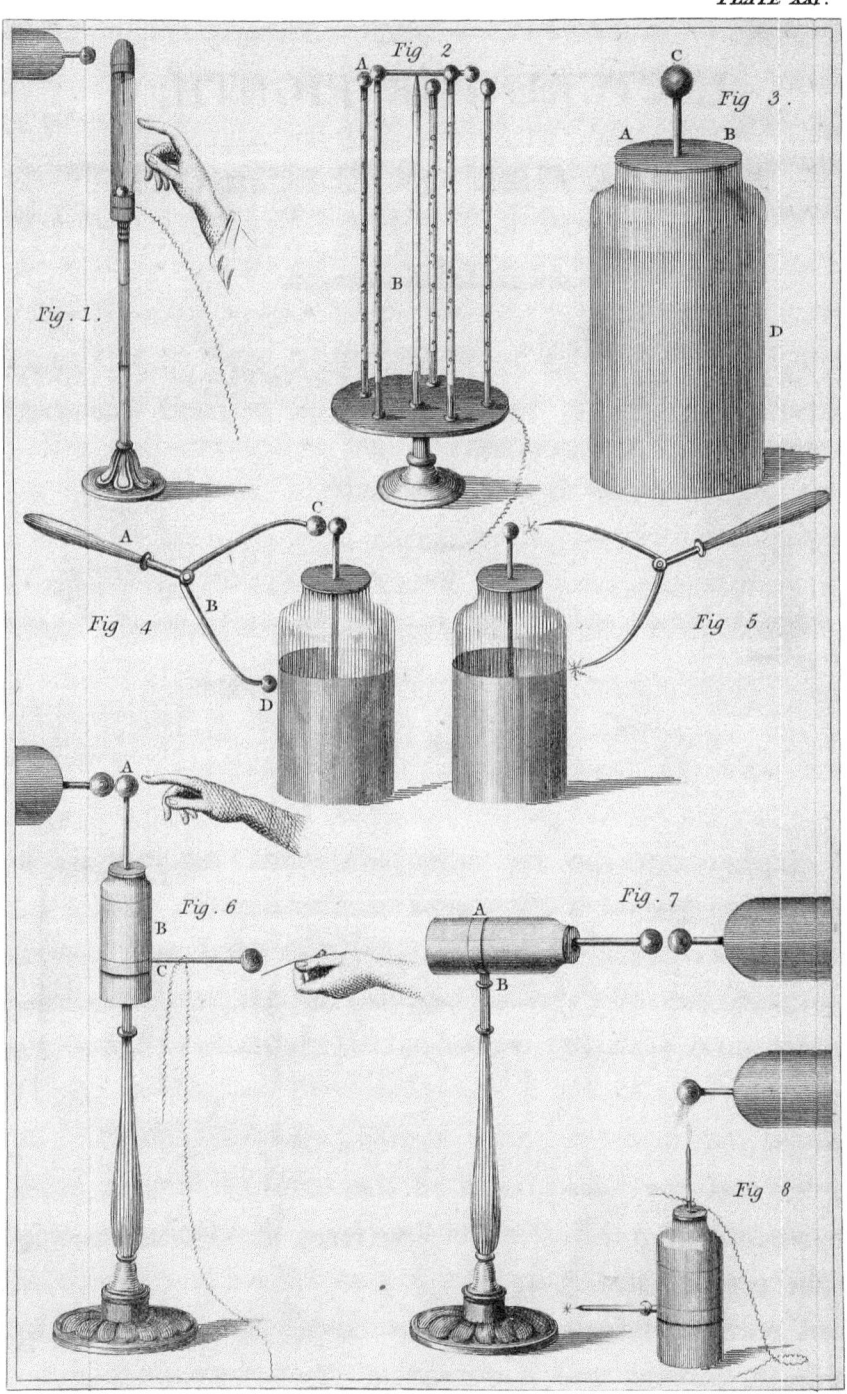

T. Noble delin.

H.Mutlow sculp.

LECTURE THE EIGHTH,

AND THE FIRST ON ELECTRICITY.

———————

THE KNOWN PROPERTIES OF ELECTRICITY EXHIBITED BY A VARIETY OF PLEASING EXPERIMENTS. THE TWO POPULAR THEORIES CONSIDERED, AND IMPARTIALLY APPRECIATED.

———

He that enlarges his curiosity after the works of nature, demonstrably multiplies his inlets to happiness.

JOHNSON.

———

NO employment can be more delightful to a rational being, than that of investigating the works of Providence. Their variety, beauty and usefulness afford the most extended and unsatiating delight—excite the most exalted admiration and astonishment—and impress the mind with the most fervent gratitude and undeviating devotion.

I never feel my importance in the scale of human beings so much as when engaged in these researches: they seem to carry me beyond the present state; by creating an ardour to discover all the charms of rectitude displayed in the dispensations of Providence, and a refined, fervent and invigorating effort to communicate these divine evidences, for the benefit of my fellow-creatures.

In the most elaborate display of the works of the great **Creator**, we find all harmoniously combining, by the simplest means, to effect the most useful and magnificent result; nothing is left either to the conflicts of contradiction, or the fluctuations of uncertainty; but all are expressive of the most exalted wisdom and benevolence, and accordingly excite our admiration, gratitude and love. Happy indeed shall I be to excite this ardour, to awaken these affections in the minds of those who have not felt these beauties of nature, aided by the extensive information and divine solace of philosophical reasoning and demonstration. With you, my dear pupils, in particular, let these communications serve to fix your principles, and induce a suitable dignity of virtue in all your actions, expressive of your sensibility to the graces of rectitude, and the precepts of Christianity; the former pourtrayed, and the latter confirmed, by every known physical phenomenon.

Of all the subjects of natural philosophy, none has been more perplexed with contrary theories, or is less understood as to its primary principles, than electricity.

In treating of this subject, I will not advert to the very ancient hypotheses which have been long exploded—such as the sympathetic powder of the peripatetics, or the effluvia emitted from and returned to excited bodies, of Gassendra and other theorists—but confine myself to the two modern theories, each of which has its adherents, and both may remain unimpeachable by experiment; though some appearances, perceivable by that test, seem to invalidate this idea of both, which I shall notice as they occur in the course of exhibiting the phenomena of electricity. I think there may be other causes producing certain effects perceived in this science, that are not yet understood by us. I have not the presumption to decide, which of the two theories alluded to is the least objectionable; therefore my judgment is unbiassed: agreeably to this determination, I shall advert to both in the course of the experiments; by which original association the possibility of either theory will be demonstrated, and that things contrary in their essential properties may produce similar results.

This mode I think the fairest; and though it will not confirm either theory individually, yet it will serve a very good purpose—that of enforcing the love and practice of justice, as the circumstances attending this subject allow me to draw an inference from them, which may be beneficial to the moral conduct of my dear pupils. Young persons are very apt to form hasty judgments; their inexperience produces this inconvenience, without their being sensible that in so doing they transgress against the laws of moral rectitude.

As the two theories of electricity serve to show us, that contrary properties may produce similar effects; let us reflect, that in human conduct also similar actions may arise from contrary causes, and that when the springs of action are unknown to us, we cannot develope their nature, and therefore must not appreciate their value: and how often do our opinions change, when facts are investigated! It is true, some actions are their own comment, and therefore cannot be mistaken; on such we may venture to decide: but others are enveloped in too much obscurity, to exhibit their real character. Appearances may be either good or bad, and yet neither originate from their primaries. This allows me the further satisfaction of guarding my pupils against the sophistry of the world, which sometimes disguises the deformity of vice under the beautiful exterior of virtue, making " the worse appear the better reason:" this perversion

of character is the more dangerous, because conveyed under the captivating form of virtue, and in the regular effective manner of reasoning, which gradually and impressively steals on the affections, and
then warps the understanding, drawing down the judgment from its
original uprightness.   Nothing tends more to shield the understanding from such attacks than the study of natural philosophy; it is
not only defensive, but also repellent.   The established principles
in the science of physics bear the form and energies of truth; and
therefore the weak blandishments and efforts of falsehood shrink
from their touch and trial, vanishing before their invulnerable
strength, and dazzled by their pure, resplendent and intrinsic graces.

In recurring to the subject of this lecture—the nature and effects
of electricity—let us, with Mr. Locke, recollect, " that the essences
" of all things are utterly unknown to us, and therefore all our pre-
" tences to discover them can be no more than mere nomination."
We can no more form a correct judgment of the essential properties
of things, without knowing their essence, than we can form a correct
judgment of human actions, without knowing the heart of man, and
the motives which actuate his conduct.

HISTORY OF ELECTRICITY.

ELECTRICITY is the sixth kind of attraction, which property
was, by Thales the Milesian, first discovered in amber six hundred
years before the Christian era.   The attractive power of electricity
was supposed by the learned, for many generations, to be peculiar to
amber; but later philosophers have discovered that this power may
be excited in different substances.

Dr. Gilbert, in 1600, made the first considerable advancement in the knowledge of electricity; he discovered that some bodies exhibited this property by friction, others by transmission only; and his communications on this subject served as a foundation to all the subsequent theories in the science of electricity. He discovered that the electric effluvium possessed both an attractive and a repulsive power—was of a subtile nature—was brought into action by friction—and that stronger evidences of it were excited in warm and dry weather than in a cold and wet state of the atmosphere.

Boyle and Otto Guericke extended their observations on the known principles of electricity. The former discovered that excited electrics, which attracted other bodies, were attracted by them in the same proportion; that feathers and different light substances were attracted by animal bodies; that an accumulated quantity of electricity would exhibit flame, and produce sound, in passing into the air; and that a conducting substance placed in an electric atmosphere, had its nature changed by that situation.

The great and comprehensive mind of Newton discovered that the electric fluid acted through the substance of glass, by evidences of its attractive effect extending to the opposite side of an excited glass tube.

Mr. Hauksbee, in 1709, invented the electric machine on the known effects of excitation. After his death no improvement was made in this science for many years; but it was revived, with much energy, by Mr. Stephen Grey, who introduced the method of retaining an accumulation of the excited power on substances by means of insulation.

Dr. Watson tried various new experiments on the electrical fluid,

and discovered that the electric spark would fire combustible substances.

In 1745 the Leyden phial was invented. In 1747 doctor Franklin, in America, communicated his observations on the science; of
which, that proving the identity of lightning and electricity was the
most important; as by that discovery he determined also the possibility of disarming the lightning of its terrors, under certain circumstances, and with certain restrictions.

Of the Affections of the electric Fluid.

Electricity is a property in substances which, when rendered
active by heat or friction, attracts light bodies; and under certain
circumstances emits a bright spark that produces a crackling noise
in passing into the air, diffusing faint sulphureous effluvia. Such
bodies as exhibit electricity by attrition are called electrics, or non-
conductors; possessing only an attractive power on the part of their
surfaces which is excited, not transmitting that power to any other
part of their substances. Conductors cannot be excited by friction;
but they suffer a diffusion of the electric fluid through their whole
substances when it has been accumulated by excited electrics, and
communicated to them.

Of the powerful effects of electricity we have very convincing
evidences, but of its essence we can say but little : however, I will
briefly relate the conjectures concerning it, which seem natural to
human conception, though, like human reason, imperfect.

The electric fluid is with good reason supposed to be an ema-
nation of fire, or light, in its subtilest form, but always combined
with other matter, which causes it to emit a faint odour.   Its action

and appearance correspond with the known properties of light and fire; and its materiality is evident, by the motion it communicates to the atmosphere in passing through it.

The two popular Theories of Electricity.

Of the two theories of electricity which have obtained the greatest share of approbation with philosophers, one is called Ellis's, being his idea as established by Volta and Mr. Atwood; the other, Franklin's, though Dr. Watson first intimated the opinions which were afterwards digested and established by Dr. Franklin. Ellis's hypothesis asserts, that what we call electricity, may be occasioned by two distinct positive and active powers: that these powers attract each other, but repel themselves: that when these powers are rendered unequal in a substance, the increased power expands into an electrical atmosphere: that the two powers exist unitedly in all bodies in their natural state, and are attracted by all substances; but that they do not become visible except when their equilibrium is disturbed, and they are separated from each other: that the friction of electrics separates their fluids, causing one to act on the surface at the place of excitation, and attract the contrary power, but to repel an excited power of the same nature with itself; which attraction of the contrary power continues till the equilibrium is restored: that in charging a jar, while one power is thrown on one side of it the other power is attracted to its other surface, from the earth, and in equal proportion.

Franklin's theory asserts—that the operations of electricity may depend on the action of one simple fluid: that this fluid repels itself, but attracts all other matter: that glass and all other electrics contain a great quantity of the electric fluid, but do not suffer it to pass through their substances from other bodies: that an excited electric has the equilibrium of the fluid broken, and parts with a portion of its natural share to a conducting body presented to it. He calls the electricity positive in a substance possessing more, and negative in one that possesses less, than its natural share of the fluid. This theory supposes that electrics never contain more than their natural share of this fluid; and hence, that when a jar is charging, as much of the fluid is thrown off from one side as is received on the other, when the jar communicates with a conducting substance; by which one side possesses more than its natural share of the fluid, and the other less; and that in forming a union of the two sides, the redundant quantity rushes to the deficient surface, to restore the equilibrium. Dr. Franklin considers the redundant power only as active, calling it positive, and the deficient side wholly inactive, or negative.

Let us turn from mere theory, to more amusing and satisfactory tests of enquiry—experiments; which we shall perceive are in general applicable to both the above theories, and it will be seen that no one can with reason wholly reject either of them.

For a considerable time after my first attempt at digesting these theories, I was induced to favour one more than the other; but after long practical experience on the effects of electricity, I find each in a certain degree supported by most of the experiments in this science, and neither absolutely confuted by any of them. For, as I before observed, there may exist causes distinct from known properties, which produce appearances and effects contradictory to both. I trust, in the course of our investigations, to justify my opinions on this subject, or at least to render my reasoning unimpeachable.

As only the attractive power of electricity was discovered in the infancy of the science, and amber the substance in which the property was more evident than in any other, the name electric, signifying amber, was adopted to express the affections of the fluid. Other bodies exhibit this attractive power, such as glass and resinous substances, silk, dry wood, &c. when rubbed or excited; but a difference is sometimes perceived between these, for an excited resinous substance generally attracts an excited vitreous one, and each repels the excited electric of the same nature with itself; which differences are expressed by the favourers of Ellis's hypothesis by the terms resinous and vitreous; while by the adherents of Franklin the former is called negative, and the latter positive.

All substances on the earth are either electrics or non-electrics. The former exhibit signs of electricity by attrition; the latter, by communication: vitreous and resinous substances belong to the former, and animal and vegetable substances to the latter, class.

The electrical Machine explained.

Fig. 1, pl. 19, exhibits an electrical machine. Electrical apparatus are constructed on various plans, and appear in many different forms; but the principles of action are the same in all, though the mode of operation varies in each of them. A, fig. 1, pl. 19, represents a glass cylinder, which is an electric or non-conducting substance. B a metallic cylinder, which is a non-electric, or a conductor of electricity. C, D, are two glass pillars, which insulate the machine, or cut off its communication with the earth, which is the grand repository of the electrical fluid. E F G H represents the frame that supports the cylinders: it is made of wood baked—a non-conducting substance.

z

At L, on the top of the pillar c, is a piece of wood, to which, at the place of its contact with the glass cylinder, a stuffed leather is attached. When the handle K is turned, the cylinder A rubs against this leather, which, by its pliability, yields to the pressure, doing no injury to the glass. The friction produced by the cylinder and leather excites the electric fluid on the surface of the glass cylinder, whence it is conveyed by the conducting metallic points M, N, to the metallic cylinder, in which it remains till some conducting substance communicating with the earth carries it thence.

The metallic cylinder or prime conductor of the machine has a chain suspended from a metal ball on the cushion c, to form a communication with the earth; otherwise no great accumulation of the fluid could be obtained, for from the earth alone we can augment this power sufficiently to render it evident to our sight. A piece of black silk (an insulating substance) is loosely attached to the glass cylinder, to prevent the dispersion of the fluid before it arrives at the points M N. Having suspended the chain from the cushion, on turning the machine and presenting your finger to the brass ball o, in the prime conductor, a bright spark will appear, and a crackling noise be heard.

Ellis supposes that the friction between the glass cylinder and rubber separates the two fluids, and causes the vitreous electricity excited by the rubber to pass to the prime conductor; and that the resinous electricity of the conductor, in its natural state, is conveyed to a ball placed on the cushion at L; and that thus the prime conductor and the ball on the cushion or rubber become possessed of contrary powers : and accordingly, when one body is electrified by the conductor, and the other by the cushion, they attract each other; but when two bodies are electrified by either of these individually, they repel each other, and attract those bodies that have less of the same power than themselves, and are still more attracted by those which possess more than their natural share of the contrary power.

According to Dr. Franklin, the effects of the electrical machine are occasioned by an accumulated quantity of the fluid produced by the friction of the cylinder; so that bodies electrified by the prime conductor, being surcharged with the fluid, on being presented to others electrified by the cushion which have less than their natural share of the same fluid, attraction takes place between them. Or if two bodies be electrified either by the prime conductor or the cushion, they repel each other, because they possess the same quantity, and therefore seek to impart their abundance to the atmosphere, or to other bodies contiguous to them that are in a natural state. Doubtless the atmosphere contains many conducting particles, and by communication may carry the effluvia to more perfect conductors, and thence to the earth.

By natural deduction we must suppose that the electric fluid is performing most important offices in the invisible world of nature, though it becomes no object of our senses till its natural state is disturbed, or its qualities are separated; for we cannot suppose that it is only our efforts that vivify its nature and create its energies.

The most beautiful and brilliant effects of the electric fluid, and the variety of electrical experiments, certainly result from disturbing the constitutional harmony of this fluid; for when that is restored, all electrical appearances vanish. Hence we may infer, that in its natural state this fluid preserves its equilibrium, performing silently its operations without constraint or undue exertion.

If the cushion of the electrical machine be made to communicate with the earth by a chain, and we turn the handle, the prime conductor will be strongly electrified with the vitreous or positive power, and strong sparks may be drawn from it; but if we unite the chain of the rubber with the prime conductor, and apply a finger, no signs of electricity will appear, the two powers or quantities being in equilibrium. The more the silk adheres to the cylinder, the stronger is the excitation. P is a screw at the bottom of the insulating pillar, to increase or diminish the pressure of the cushion against the glass cylinder.

You must always remember, that when positive or vitreous electricity is required, a chain should be suspended from the rubber to communicate with the earth; but when resinous or negative electricity is to be produced, the cushion must be insulated, and the prime conductor communicate with the earth by a chain, and the fluid taken from the ball on the cushion.

That electrics or non-conductors have only that part of their surfaces which is in immediate contact with the exciting cause rendered electrical by excitation, is evident by rubbing a glass tube with black silk; when the part rubbed will attract a feather, but the remainder of the tube will not. That non-electrics or conductors, when receiving the fluid from excited electrics, exhibit signs of electricity in every part of their surfaces, is proved by presenting your finger to the prime conductor while the machine is turning; for then a spark will be drawn from any part of it; showing the universal diffusion of the power on conducting substances.

May not the evidences of electricity, on excited substances, be partly independent of the electrics, and external; as in the atmosphere for instance, at the place of excitation, which, on the motion communicated to it by the attrition, may collect the electric effluvia; and the electrics themselves only rendering the accumulations local, by not transmitting the electric fluid through their pores? May not also the difference of the effects sometimes perceived in the excitation produced by means of glass, and that obtained by wax, be dependent on different surfaces, or the state of fire in these substances, which may cause them to attract different particles of the atmosphere?

We know that the atmosphere is composed of the greatest possible variety of heterogeneous particles. May not some of these have affinity with particular kinds of bodies, and others with bodies of a different description, and produce either a different quantity or quality, of such particles as impart that quality or quantity to the atmosphere, at the place of excitation, or induce a stronger or a weaker portion of the attractive particles in their united form?

Agreeably to my plan laid down at the beginning of this lecture, I shall use the terms vitreous or positive, resinous or negative, conjointly; the two former, when speaking of the electricity produced from excited glass, as from the electrical machine, fig. 1, when the cushion communicates with the earth by a chain, and the fluid is taken from the prime conductor; and the terms resinous or negative, when the prime conductor communicates with the earth, and the fluid is procured from the cushion.

Our sight, hearing, smell and feeling, all bear evidence of the materiality of the electric fluid. For it appears like fire; it smells sulphureous; strikes the body it touches; and, by giving motion to the atmosphere in passing through it, produces sound.

Experiment.
The agitation caused in the air by the electric fluid is rendered

evident by placing a pointed brass wire, A, fig. 2, in the prime conductor, and holding the hand over it, when a strong current of air
is felt.

Of electrical Attraction and Repulsion.

I have placed on the prime conductor of the electrical machine
a figure, holding a brass wire, to which are suspended some downy
feathers; on exciting the machine these become possessed of vitreous
or positive electricity, and diverge, either because they are possessed
of the same sort or quantity of electricity, or perhaps because the
currents of their issuing fluid, acting in opposition to each other, drive
them asunder. Lay your finger on the prime conductor, and the
plumage will fall, the equilibrium of the fluid being restored by your
communication with the earth. Letting the feathers diverge again,
and presenting a finger to them, they fly towards it, because the
hand is possessed of either a contrary quality or less accumulated
power of electricity. Suspend a pair of pith balls from the point A,
in the insulating stand, fig. 3, pl. 19, and electrify them positively or
vitreously, by the prime conductor; when they will repel each other.
Then holding another pair in your hand, and presenting these to the
insulated electrified ones, the two pairs possessing contrary powers
or qualities will approach each other, as seen fig. 4. Thus we perceive that the effects of electricity become evident to us only when
the equilibrium is destroyed, whatever that be, whether of quantity
or quality. A body immersed in an electrified atmosphere becomes
possessed of a contrary power or quantity to that of the electrified
body. May not this effect be thus produced?—Immersed in an
electrical atmosphere, it attracts the same quantity, or that quality
of electricity to itself, and as soon as it has so done, becomes possessed of the same quantity or quality, and is therefore by the
issuing fluid divided from the electrified body, as before intimated.
This effect, let the cause be what it may, is important to all our ob-

servations and experiments in electricity, as, without considering it, we cannot understand their results.

Light Bodies convey the electric Fluid.

Experiment.

If a feather suspended by a piece of common thread be held to the prime conductor, it will be attracted so long as the machine is turning; which shews that it conveys the electrical fluid to the hand as soon as it receives it; for if the feather became charged with the fluid, it would remain separated from the prime conductor.

Experiment.

Two pith balls suspended from the wire in the prime conductor repel each other, as seen by fig. 5; because they become possessed of the same power or quality, and being charged with the fluid which cannot be drawn off by a good conducting substance, they are separated from each other, either by the intervention of their atmospheres, or by the natural repellency of their electricity. Presenting either glass or wax to the electrified balls does not alter their state. —Does not this favour my idea, that the difference perceived in the effects of electricity may not reside in the vitreous or resinous electric, but may depend on the different action and accumulation of the particles of the atmosphere on their various surfaces?

Experiment.

To know the extent of an electrical atmosphere at any time (for this varies according to the accumulated power), hold a couple of pith balls within two feet distance from the machine, and if the excitation be very powerful they will separate from each other; but if it be weak, this effect takes place only when they are held very near to the apparatus.   By this experiment we determine the extent of the electrical atmosphere of the machine, or of any other electrified body, at a given time.

Experiment.

The issuing current of electricity is shewn by a very pleasing and simple experiment with a brass cross, placed on a pointed wire in the prime conductor, as exhibited by fig. 6: for on turning the handle, the cross revolves, and each of its terminating points and bent parts emits a spark.   The revolution of the cross arises from the issuing fluid being repelled by the air, agreeably to the laws of action and reaction; for the cross would not revolve if the points were not bent.   For if the flies were straight tubes, issuing from a hollow cross stem, and the whole filling constantly with water, the fluid would jet out at the extremities of the cross, but the fly would not revolve, because the action would be directly from the centre of the cross, and the air would react in the opposite direction to that only.   But if the hollow tubes were all bent one way, there would be a pressure against the bent parts of them by the air, which causing a reaction of the fluid in the levers or arms of the cross, the water would not only flow out, but the cross revolve.   The motion of the fly by electricity is produced in the same manner.   For the fluid issues from the centre of the cross, and being thrown off at the points, the steam is resisted by the air, and thus the cross revolves in a contrary direction to the issuing current.

The attractive and repulsive effects of the electric fluid are very pleasingly exhibited by a double peal of bells suspended on an insulating stand; fig. 1, pl. 20. The chains, *a, a,* from the centre bells, A, A, communicating with the earth through the tube, and the support B B of the other bells being insulated, on turning the electrical machine the electricity is communicated to c c c c, and thence to the balls, which by their different state of electricity, or by the issuing current are driven towards the bells A A, then again they return to c c c c, and so on, and thus produce a peal by their vibratory motion. Tie ribbons to the chains at D D, and raise them from the table, when the ringing will cease: May we not suppose this happens because no conductor communicating with the earth offers itself?

Experiment.

Holding your hand, as represented fig. 2, within a short distance of the prime conductor, and laying on it a piece of cotton, this will move with great velocity backwards and forwards between your hand and the metallic cylinder, stretching out in the direction from one to the other, conveying the fluid to the earth, or exchanging powers with it.

Experiment.

Charge a tumbler with the electric fluid, by inverting it over a pointed wire in the prime conductor; then placing it over pith balls on the table, these will convey the fluid, rising and falling, as implied by fig. 3. When the equilibrium of power or quantity is nearly restored, the motion of the balls will be slow, and at length entirely cease. Sometimes the table is too dry to receive the fluid from the balls; when this is the case, on touching the glass with the hand or a piece of metal, both good conductors, the balls will again begin to perform their vibrations.

A very popular and diverting experiment, on the attractive and repulsive appearances of electricity, may be exhibited by means of two brass plates, one suspended from the prime conductor, and the other lying on a brass stand.

Experiment.

Cut neatly some figures, as dancing, in tin foil, or thin leaf-gold, and place them on the lower plate, which must be about four inches distant from the upper one, as seen in fig 4.   Then bring the peal of bells also to the brass wire; and on turning the machine the figures will dance between the plates apparently to the music of the bells. If you lay some grains of gold or silver dust on the lower plate, these, rising and falling at the same time with the dancing figures, in the dark will appear like a shower of fire: on touching the prime conductor with your finger, the dance will cease, as if at the word of command.

Experiment.

Fig. 5, pl. 20, represents a quadrant electrometer, which shows the accumulation of electricity, or power of the machine, by repulsion.   For when the machine is most powerful, it flies to the greatest possible distance, that is, to the line perpendicular to it, or to 90°, either by the repulsive quality of the electricity or by the extent of its atmosphere.

Experiment.

That a conducting substance attracts an electrified body as well as the latter does the former, is rendered evident, by suspending a small brass bucket of water to the prime conductor; when on turning the machine, and presenting your finger to the water, it rises in the form of a cone towards your finger.

Experiment.

The bucket, fig. 6, has a capillary tube, through which you will perceive the electricity causes the water to flow in a stream; on stopping the motion of the machine, the water drops but slowly from the tube. This shows that electricity quickens the motion of fluids; from which we may infer, that the circulation of the fluids in animal bodies may be accelerated by electricity, and consequently that the blood may at all times be affected by it.

Experiment.

When a person is to exhibit the electric spark, he must be stationed on an insulating substance; a glass-footed stool, which is represented by A B C D, fig. 7, answers the purpose. The top of this stool is of baked wood with the corners rounded, because pointed bodies facilitate the dispersion of the electric fluid. May not this effect arise from such bodies parting from a very small quantity of the fluid at a time, and therefore being less impeded by the air? When we present a finger to the ball on the prime conductor, a strong broad spark appears, and a sound is heard; but when we present a finger to a pointed body placed on the conductor, the spark and sound are both much less observable.

Experiment.

Let fig. 7 represent a person electrifying by the machine, standing on the stool, and holding a chain suspended from the prime conductor, no other body communicating with her; on touching the electrified person sparks will appear: if the person electrified hold a brass wire and ball, and present the latter toward another person, the spark will appear more vivid, and the conducting person will receive a sharper sensation than the electrified one, because the surface of the ball is larger than that of the wire, and therefore imparts a greater portion of the fluid at the same instant of time.

Experiment.

A diverging bluish flame, visible in the dark, will issue from a pointed body held in the insulated person's hand, if no one touch her; but as soon as she is touched the flame will disappear. This proves that pointed bodies readily throw off the accumulated electricity.

Experiment.

Presenting a piece of leaf-gold or tin foil, suppose cut out in the form of a bird, see fig. 8, to the electrified person, it will immediately fly towards her; when possessed of her electricity, it will be repelled: if then received again by another person, it will discharge itself, and continue to fly backwards and forwards, forming a very pleasing experiment.

Experiment.

That the electric matter is fire, or in the most intimate union with it, is evident by its effects on spirits of wine, which it causes to flame. Let the electrified person hold a spoon containing some warm spirits of wine, and another present the tip of her finger or the brass point to it, and it will immediately emit a spark which will fire the spirits.

Experiment.

If two persons be electrified by communication with the prime conductor at the same time; on either touching the other, no spark will appear, because they are both equally charged with the electric fluid: that they are so, may be ascertained by their holding a pair of pith balls, suspended by threads, in their hands, which will repel each other.

Experiment.

The electric matter diverges considerably more in vacuo than in the open air. The tall receiver, fig. 1, pl. 21, is exhausted of air: on placing it in contact with the ball of the prime conductor, the fluid issues from the point A, fig. 14, and flies towards the hand, filling the whole receiver. If we touch any part of the glass, the light will be attracted towards the finger that touches it, being the nearest conducting substance within its sphere of attraction. We hear no explosion in a vacuum, because the impressions of the atmosphere are removed.

Experiment.

Electricity will enable us to form a beautiful representation of the stars. Paste small pieces of tin foil on a piece of glass, cutting the edges between the pieces to resemble stars. Placing the glass in a wooden frame, with a brass ball at one end; on connecting the other extremity by another ball communicating with the prime conductor, the intervals will be illuminated, and a brilliant constellation will appear. If the whole of the glass, excepting the stars, be made black, the appearance will be still better. Any other form may be exhibited in the same way, by leaving intervals for the fluid between the conducting substance that forms the figure. Receiving the fluid from the conductor, while the other extremity of the wire communicates with the cushion of the machine, renders this representation in the highest degree splendid.

Experiment.

Fig. 2, pl. 21, represents spiral tubes. Round these tubes is placed tin foil cut in small pieces, with intervals left between them, to shew the passage of the fluid. Brass caps are fitted to these tubes, to receive the fluid from the electrical machine, by the revolution of a brass balance, A, which alternately touches them. When one of the balls at the extremity of the wire receives the fluid, it communicates it to the next, thence it goes to the next, and so on; an interchange of qualities taking place. The balance wire is supported on an insulating pillar, B. The bottom of the stand has a wire communicating with all the tubes, and forming a communication between them and the earth.

Experiment.

The northern light, or aurora borealis, is beautifully exhibited by a vacuum in a tube about two feet long. On presenting the tube to the conductor, the whole of it becomes illuminated, and continues so for some time, even when taken from the electrical machine. If it be drawn through the hand, the light will become much brighter, and more dense in appearance. But this operation discharges the fluid; yet it will flash at intervals, like the northern lights, if the hand be applied to it. This effect will continue many hours if the tube have received a large quantity of fluid from the conductor.

Having exhibited some of the most popular and pleasing experiments in the science of electricity, I shall defer an investigation of its more important and grand effects; leaving you time to meditate on the wonders already displayed of this powerful agent. Equally surprising are all the agents employed by Divine Providence to effect what we call the order of nature, with those we have been contemplating in this course; but, being familiar, they do not excite the same astonishment.—

What prodigies can Power Divine perform
More grand than it produces year by year?
And all in sight of inattentive man!
Familiar with the effect, he slights the cause.   COWPER.

But the attentive man sees God in all ;
Adores him in each object ; inspired, his
Will obeys.   Charmed with goodness infinite,
Gratitude warms his heart ; refines his nature ;
Exalts his virtues ; and attunes his ardent soul
To joy, love and praise !—celestial harmony !

PLATE XXII.

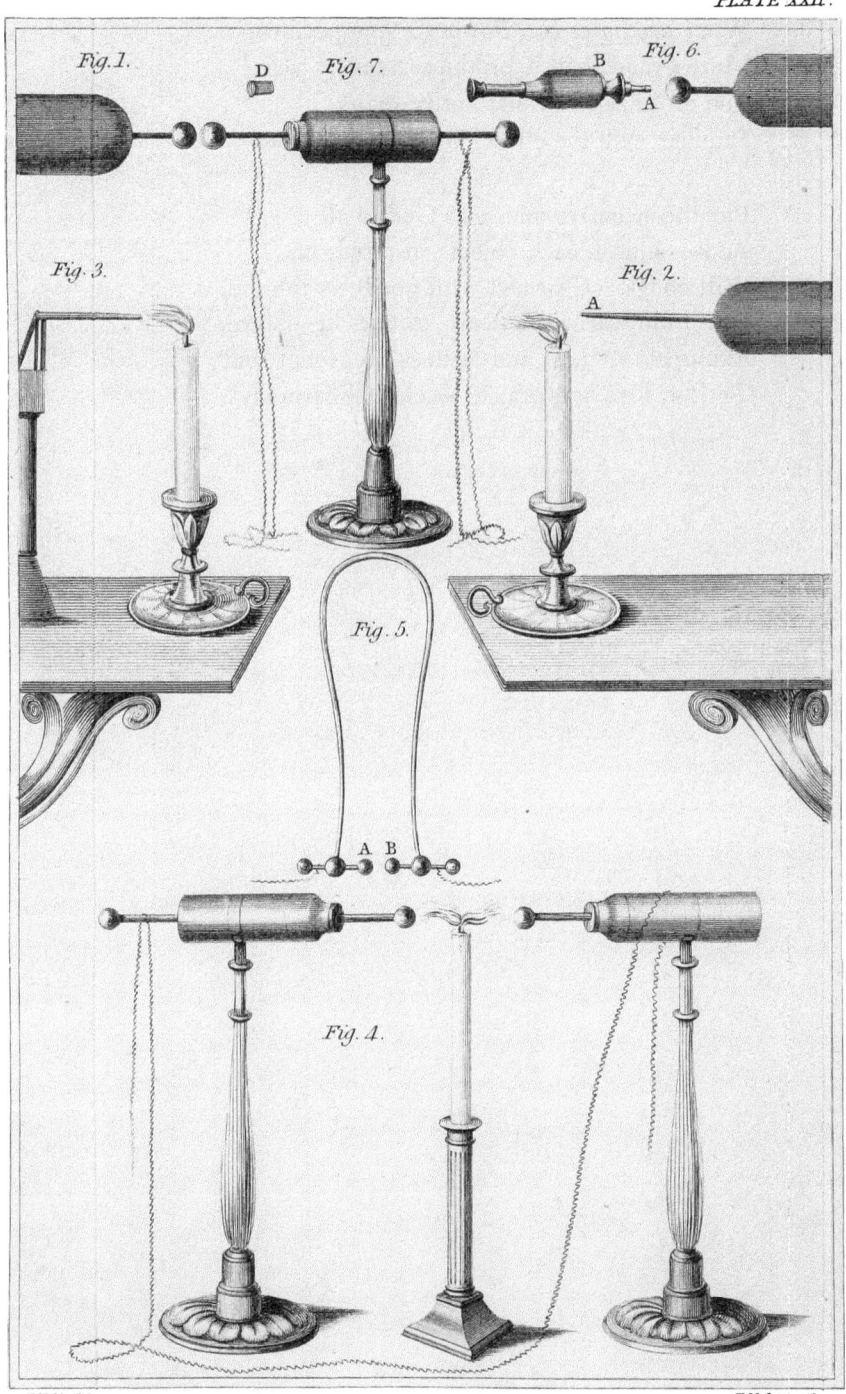

Fig. 1.   Fig. 7.   Fig. 6.
Fig. 3.   Fig. 2.
Fig. 5.
Fig. 4.

T. Noble delin.   H. Mutlow sculp.

PLATE XXIII.

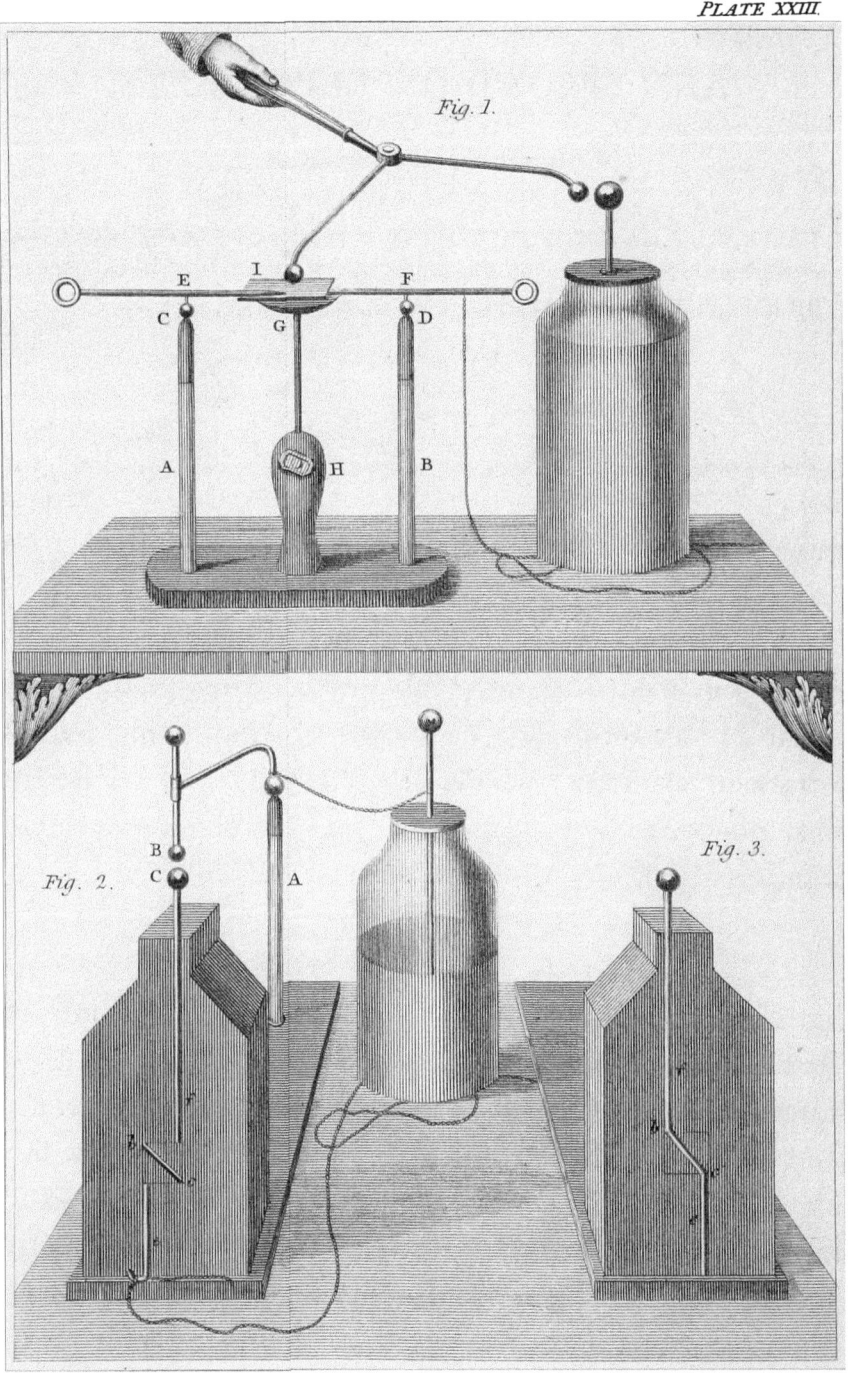

Fig. 1.

Fig. 2.

Fig. 3.

T. Noble delin.ᵗ

H. Mutlow sculp.

LECTURE THE NINTH,

AND SECOND ON ELECTRICITY.

OF THE LEYDEN JAR.—OF THE DIFFERENT EFFECTS OF ELECTRICITY PERCEIVED ON POINTS AND KNOBS.—THE PROPERTIES OF CONDUCTORS CONSIDERED.— OF THE IDENTITY OF LIGHTNING AND ELECTRICITY.

The first and greatest work of a Philosopher, is to try and distinguish appearances; and to admit none untried.   EPICTETUS.

WE will now continue our researches in the investigation of the subject of the electrical effluvium, and contemplate its amazing energies when accumulated, by means of conducting particles, on the surface of electrics; which, by its variety and connection with the other phenomena of nature, will include much entertaining and useful information.

Of the Leyden Jar or Phial.

Fig. 3, pl. 21, represents the Leyden jar. It is made of glass, and coated, both exteriorly and interiorly, with tin foil (a conducting substance), to convey the electrical fluid to all parts of the surface, on which the excited electricity of the machine is to be accumulated and retained. The jar has a stopper, A B, of baked wood. In the centre of this stopper a strong brass wire is placed, with a ball, c, at top. The other extremity of this wire communicates with the inside coating of the jar, to prevent the powers or qualities of electricity from uniting. The coatings of the phial do not rise

higher than within two inches of its top.   When the ball c is removed, a point appears; and when this is unscrewed from the stopper, the electrometer, fig. 5, pl. 20, may be put on in its place.   This
is used to ascertain the power of a charge received by the phial.

To charge the Leyden Phial.

On turning the machine, and placing the ball A in contact with
one in the prime conductor, strong sparks will pass between them;
and by means of the brass wire, and inside coating of the phial, the
electric fluid from the machine will be accumulated on the inner
surface of the glass; for while the outer surface of the phial communicates with the earth by any conducting substance, an accumulation takes place.   While the fluid is thrown on one side of the
phial, an equal quantity of it is either attracted to the outer surface
of a different quality, or the same quantity is thrown from it, and
thus the two surfaces are in different states.   When the phial is
charged, on applying one hand to the ball c, communicating with
the positive or vitreous electricity, and the other hand to the coating
D, a sudden and unpleasant shock is felt, produced by the union of
the powers or qualities.   This sensation is more strongly felt in
the breast than in any other part, and if violent, may injure the
animal functions; therefore we use the discharging rod, exhibited
fig. 4, pl. 21.   A is a glass handle to insulate the body; B a brass
wire terminating in points.   This instrument, on which the balls c D
are occasionally screwed, being applied in the abovementioned manner
of discharging a phial, no sensation is felt, as the fluid passes from
the points or balls through the conducting substance without communicating with the body.

As glass is impervious to the electric fluid, any form of it may be
charged, when the two surfaces are separated from each other.   The

effects of charged glass were discovered in 1745, by Cunæus, a native of Leyden, from which local circumstance the jar derives its name.

In the history of electricity Professor Muschenbroek appears very conspicuously, having much improved the science by his investigations. He discovered that electrified substances exposed to the atmosphere soon lost their excited electricity, from which he inferred, that if the communication with the atmosphere were cut off by closing the surface, and the fluid accumulated by means of a conducting power, the effects of electricity would remain a longer time. His first experiment was made with water confined in a glass bottle, which confirmed his opinion, and surprised him by its evidences. Suspending this bottle by its inside from the prime conductor, and charging it highly, on applying one hand to the brass wire at top, and the other to the outer surface of the bottle, he felt a severe shock in his arms and breast.

This discovery was improved by Gralah; who further increased the strength of the charge, by connecting several jars together, and charging them at the same time: this he called the electrical battery. This gentleman likewise discovered, that the charge of a small jar would communicate a shock to several persons at the same time. The residue of a charge was also first perceived by him.

Dr. Watson prosecuted the observations before made in this science. He contemplated the effects of the electric fluid on a glass phial coated with water, and perceived that the charge was considerably increased when the water was made hot. I believe he was the first who formed the coating of jars with metal, knowing it to be the most perfect conductor: and the trial justified his ideas, for he found the explosion from these jars was greater, than from those charged by means of water.

Canton observed, that when a jar was insulated on both sides, on receiving the fluid from the conductor, though no explosion followed, sparks could be taken from the outside and inside alternately, till the equilibrium was restored.

M. Monnier discovered, that during frost the charged glass contained the electric fluid much longer than in wet or damp weather. Probably this, added to the stronger impressions of the air in that state of it, may cause our nerves to be more braced, and our spirits more invigorated, in a frosty than in a damp state of the atmosphere.

According to the learned Dr. Franklin, in charging the jar, as much of the fluid is thrown off from one surface as is received on the other: he supposes that there is no more of the fluid in the glass after charging it than there was before; and thinks that glass is in its natural state as full as possible of the fluid, and therefore contains no more when charged than before. In discharging a jar, he conceives one power only as active, namely, the redundant or positive; and in support of his opinions produced the following experiment.

Experiment.

Holding a thread near the coating of a jar while the inside was receiving the fluid from the prime conductor—as represented fig. 6, pl. 21—he touched the ball at top with one finger; and as he drew sparks through it from the inside, he perceived the thread attracted by the outer coating, conveying, as he thought, the fluid received by him, through his body to the deficient surface B C; from which it was propelled in charging the inside one. But I think the same appearances might arise from the union of two different fluids.

The ingenious Mr. Ellis affirms, that in charging a jar, the two surfaces are in contrary states, and possessed of contrary natures ; that the outside attracts as much of the resinous electrial fluid on its surface, as it receives of the vitreous power from the conductor ; and that this is evident by the strong attraction between the two sorts of electricity. In discharging the jar, he conceives the two powers rush into union with mutual activity. To support his idea of both powers being equally active, he produces the experiment of passing the charge of a jar through a card, as I shall show in a subsequent part of this lecture ; in which the perforation certainly exhibits the appearance of a double current.

The idea of the coating of a jar not containing the charge, but only acting as a distributor of the electrical fluid on the surface of glass, was communicated by Dr. Franklin ; being the result of an experiment undertaken by him to ascertain what part of a Leyden jar received the charge. He charged a jar and removed the stopper ; in this he perceived no virtue. He then placed one hand on the coating, and the other in the water inside the jar ; when he felt a smart percussion. Charging the phial again, he poured out the water into an empty insulated jar, in which, on applying his hands, he perceived no charge. From these results he inferred, that the charge must have been dependent on the glass, and perhaps within its substance.

May not this experiment sanction my idea of the accumulation being only on the surface of the glass, or in the atmosphere immediately contiguous to that surface ? The charge was not found in the water; and if the glass be impervious to the fluid, how can the accumulated power pass into it?

The most expeditious mode of charging a jar, is by letting one surface communicate with the rubber of the machine, and the other receive the fluid from the prime conductor.

Experiment.

In discharging the jar, if we use the points of the discharging rod which present themselves on unscrewing the balls, as represented fig. 7, pl. 21, we shall perceive the fluid passing off gradually without an explosion; and that on the point presented to the resinous or negative surface, on the outside of the jar, the fluid will appear like a diverging stream of rays (see fig. 7, pl. 21); but on the one presented to the ball at top of the jar, a star only will be observed.

From the following experiments, as well as by those I have already exhibited, I hope to convince you that these effects may arise either from two different powers of the same fluid, or from two fluids of a contrary nature. For in the latter as well as in the former postulatum, the effects may be weaker in one part than in the other; the same as if one were redundant and the other deficient; which added to the velocity of this fluid, may make both appear active on discharging a jar, though only one may actually be so; and also that both may be active, though we cannot, from the velocity of the motion, perceive where the two fluids unite. Doubtless if one be weaker than the other, the weaker power will not act so powerfully, or extend in its effects so far as the stronger one; and this may occasion the appearance of a star on the point negatively electrified; and by the superior force of the positive or vitreous power, extending itself to a greater distance, it may compress the resinous or weaker, so that the place of union may be nearer to the latter than to the former. I merely offer these conjectures; if they can be confuted by demonstration, I am open to such conviction; but till then I shall think myself justified in retaining them. The experiment in vacuo, exhibited in the last lecture, supports my opinion; for the positive power flies towards the hand, to any part where it is placed; and appears to form its union nearer to the resinous or negative than to the vitreous or positive power.

No charge can be received by a jar when both sides are insulated, or when both sides communicate with the earth.

Experiment.

Screw the bottle, fig. 7, pl. 21, with its belt A B, on the insulated stand, placing the ball communicating with the inside of the bottle in contact with the prime conductor: then on applying the discharger no explosion will be heard, the jar not being charged, because its communication with the earth was cut off on both sides.

Experiment.

On the other hand, forming a communication with the earth on both sides, by suspending a chain from the ball at the top of the bottle, and from that on its coating, as seen in fig. 7, pl. 22, and taking the fluid from the prime conductor, fig. 1, pl. 22, on applying the discharger, no explosion is heard, the circuit being complete.

Experiment.

That the coating a jar with a conducting substance, merely answers the purpose of conveying the fluid to all parts of the surface, and that the charge does not remain in the conducting particles, is proved by experiment. Charge a pane of glass, by laying a piece of loose coating on it, and connect it with the positive conductor; then on striking off the coating, and immediately applying one finger to the upper surface of the glass, and a pointed wire to the opposite one, a spark will be perceived, and in the dark a diverging stream of light will appear. The coating is necessary to diffuse the fluid; for without that addition the effect would be partial and evanescent. The effort excited by the divided fluid, or fluids, to unite, is evident in a jar very highly charged; for we perceive, at the upper edges of the coatings, rays of light darting from the different qualities or powers soliciting a union, and which sometimes they obtain and discharge the jar.

It is not necessary to charge a jar very highly for all experiments, and it is absolutely requisite we should be cautious on this head; for though glass does not suffer the union of the two surfaces through its substance, when moderately charged, yet sometimes a charge will be so great as to cause a separation of its particles whilst charging; and this may arise from a single force only acting in one direction, or by the united compression of two forces greatly augmented. It is safer to apply the discharger twice to a jar, because sometimes there will be the residue of a charge remaining after an explosion.

Experiment.

The electrical shock may be communicated to many persons, by one of them placing a finger on the coating of a charged jar, and another person touching the ball at top, after having formed a chain by the union of hands. If, instead of touching each other, they use brass balls to unite them, leaving intervals between these, a bright spark will appear at each ball on discharging the jar. The shock of charged glass is instantaneous. Experiments have shewn that the motion of the electrical fluid is so rapid, that in observing its effects at several miles distance, no interval of time is perceived.

The electrical fluid always endeavours to place itself in equilibrium by the shortest way; as is evident by the experiment of passing the shock of a charged jar through a particular part of the body, without communicating it to any other. For the whole body is equally a conductor of the fluid; but the discharging rod being applied to two particular parts of its surface, the communication is formed by a shorter way than through the whole body.

Experiment.

Place a pointed wire on the top and another in the belt of the phial, as shewn fig. 8, pl. 21; and when it is charging, sparks will be seen in the dark from both points; the one at top exhibiting a diverging stream of rays, and the one on the side a star; as if the outer surface was receiving the quantity of electrical fluid thrown from the inner one. Yet it is impossible to say that it actually is so—for the reception of the fluid on one point might exhibit the same appearance as the rejection of it does on the other.

Experiment.

That the two sides of a Leyden jar are in contrary states, we perceive by screwing into a jar two wires, with balls; when, on throwing the fluid on the inside of the jar, and holding a light pith ball between the brass ones, the former vibrates between the latter, because the surfaces are possessed of contrary powers or qualities.

Experiment.

If we present the outside coating of a jar to the prime conductor, and connect the ball at top with the earth, on applying a pith ball, electrified by an excited glass tube, we shall perceive that the outside of the jar possesses positive, or vitreous, and the inside negative, or resinous, electricity.

Experiment.

Place a long wire, like A, fig. 2, pl. 22, in the positive, or vitreous, conductor of the machine; then turn the handle, and hold the flame of a taper near the ball, and it will be driven forward by the electrical fluid.

On fixing the same wire in the cushion, and presenting the taper to it whilst the machine is turning, the flame is drawn towards the point, as seen at fig. 3. Does not this appear like an electrical atmosphere surrounding the machine, and the fluid circulating in it?

Experiment.

The following experiment exhibits the conducting power of flame. Place a ball on the outside, and one on the inside of a jar, and charge the inside positively; then present the flame of a taper between the balls, and it will be drawn towards both of them. This will be better seen by charging one jar positively and another negatively; and then introducing the flame between the balls, which should be placed at about four inches from each other. The opposite and uniform direction of the flame is exhibited by fig. 4, pl. 22.

Experiment.

It is said that the conducting principle in animal bodies is lost after death; if so, it appears that vitality has an affinity with the electrical fluid. I have mentioned that the electrical fluid always endeavours to place itself in equilibrium by the shortest way; an experiment sometimes exhibited on this subject not only serves to confirm this idea, but also that of the electrical atmosphere encompassing bodies, and which I suppose may cause the apparent repulsion of the same sort or power of electricity, but which is overcome by the attraction of the stronger power of this fluid in one of two light bodies possessed of different powers, or by the strong attraction between the two different qualities of electricity.

Experiment.

If two balls, fixed at the end of a bent wire ten feet long, as exhibited fig. 5, pl. 22, be brought within half an inch of each other, on endeavouring to discharge a battery through the brass wire the fire passes through the air between the two balls A B, in preference to going through the wire. The atmosphere of the balls must be strongly impregnated with the effluvia, for the wire will be melted to a certain distance from them; which we may suppose arises from the atmosphere of the fluid, as the fluid itself passes another way.

The electrical battery being attended with danger, I never use one; and every experiment essential to accurate information in this science may be performed without it. Yet as batteries are frequently used by electrical experimentalists, I shall just mention how they may be used with probable safety Connect one end of the discharging rod to the end of a long slender wire communicating with the positive side of a battery, and apply the other end of it at a little distance from the bottom of the jars. Use the points only of the discharger for this performance.

In the new abridgment of the Philosophical Transactions, published by Messrs. Baldwin, the following curious experiments, communicated by Mr. Stephen Gray, are introduced, page 490, vol. vi. p. 3.

" I have often observed in the electrical experiments made with a glass tube and a down feather tied to the end of a small stick, that after its fibres had been drawn towards the tube, when that has been withdrawn, most of them would be drawn to the stick as if it had been an electric body, or as if there had been some electricity communicated to the stick or feather; this put me upon thinking, whether if a feather were drawn through my fingers, it might not produce the same effect, by acquiring some degree of electricity. This succeeded accordingly on my first trial; the small downy fibres of the feathers next the quill, being drawn by my finger when held near it: and sometimes the upper part of the feather, with its stem, would be attracted also, but not always with the same success. I next tried threads of silk of several colours and sizes, which I found to be all electrical.

" Having succeeded so well in these, I proceeded to larger quantities of the same materials, a species of ribbon both of coarse and fine silk of several colours; and found that by taking a piece of either of these of about half a yard long, and holding the end in one hand, and drawing it through my other hand between my thumb and fingers, it would acquire an electricity; so that if the hand were held near its lower end, it would be attracted by it at the distance of five or six inches.

" After this I tried several other bodies, as linen of several sorts, viz. Holland, muslin, &c. and woollen, as of several sorts of cloth, and other stuffs of the same materials. From these I proceeded to paper, both white and brown, finding them, after they had been well heated before rubbing, to emit copiously their electric effluvia. The next body that I found the same property in, was thin shavings of wood; I have only as yet tried the fir shavings, which are strongly electrical.

" All these bodies will not only by their electricity be drawn to the hand, or any other solid body that is near them, but they will, as other electric bodies do, draw all small bodies to them, and that to the distance sometimes of eight or ten inches. Heating them by the fire before rubbing, very much increases their force.

" There is another property in some of these bodies, which is common to glass; that when they are rubbed in the dark there is a light follows the fingers through which they are drawn; this holds both in silk and linen, but is strongest in pieces of white pressing papers, which are much the same with card-paper; this not only yields a light as above, but when the fingers are held near it, there proceeds a light from them, with a crackling noise like that produced by a glass tube, though not at so great a distance from the fingers. To perform this, the paper before rubbing must be heated as hot as the fingers can well bear.

" A down feather being tied to the end of a fine thread of raw silk, and the other end to a small stick, which was fixed to a foot that it might stand upright on the table; there was taken a piece of brown paper, which by the foregoing method was made to be strongly electrical; which being held near the feather, it came to the paper, and I carried it with the same till it came near the perpendicular of the stick; then lifting up my hand till the paper was got beyond the feather, the thread was extended and stood upright in the air as if it had been a piece of wire, though the feather was near an inch distant from the paper. If the finger were held near the feather in this position, the greatest part of the fibres next the paper would be repelled; when at the same time if a finger were held to the fibres that were remote from the paper, they would be drawn by it.

" I repeated this experiment without the feather, viz. by a single thread of silk only, of about five or six inches long, which was made to stand extended upright as above mentioned, without touching the paper; then placing my finger near the end, it would avoid, or was repelled by it; but on placing my finger at about the same distance from a part of the thread, that was about two inches from the end, it was attracted by it."

By the experiment with inflammable air in a pistol, as well as by the effect we produced on spirits of wine in my last lecture, we perceive how easily electricity acts on inflammable bodies, even so as to ignite them.

The pistol, fig. 6, pl. 22, is made of brass; it incloses a glass tube, with a wire passing through it, having one knob at A, which is exposed to receive the spark, and another at B, communicating with the barrel of the pistol. To fire this pistol, we admit inflammable air in the barrel, by holding it downwards, thus letting the lighter air ascend into it. This manner of charging is performed in half a minute, for some of the atmospherical air must remain. Before reversing the pistol put in a cork, to prevent the escape of the lighter air. On applying the knob A to the ball in the prime conductor of the machine, to receive the electric sparks, these will pass to the barrel, firing the air within it, which exploding, the cork D is projected with violence. The glass tube prevents the dispersion of the sparks in any other direction than in the required one. The easiest and best mode of charging the pistol, is by means of one drop of strong ether, which remaining corked up in the pistol for one hour, will explode on receiving the sparks. The other mode is by a preparation of iron filings, put into a glass vessel with a small quantity of vitriolic acid, and about ten times their weight of water; which mixture produces an evaporating inflammable gas.

Fig 1, pl. 23, represents the universal discharger. On its base are fixed two perpendicular glass pillars, A, B; to the top of each is cemented a brass cap, C and D, which turn on joints in any required direction, and carry brass sliding wires, E, F. The table G rises or falls according to the height required, and is fixed in its place by a screw at H.

After charging the Leyden jar, and placing a card, I, on the table of the universal discharger, I lay the point of one wire under the card, and place the point of the other on the top of it; then connecting the coating of the charged jar with the point, under the card. To discharge the jar through it, I place one knob of the discharging rod on the ball at the top of the bottle, and the other on the point of the wire, on the top of the card; and then the union of the two sides of the jar will be completed by the electricity passing through the card. The hole in the card, after the operation is performed, will show that the electricity passed through it.

The manner of the perforation in the card certainly has the appearance of a double current; yet I think it may be produced merely by the great attraction of the positive or vitreous power causing a union independent as it were of the individual action of the negative one.

Experiment.

I will render a bottle of water luminous; to show that water is a conductor of electricity, or at least suffers a free passage of the electric fluid. Hanging a chain from the outside coating of a charged jar, and suspending the other end of it in a bottle of water; then laying another chain under the latter at about a quarter of an inch from the former one, and placing one knob of the discharger on that end of the chain which is not connected with the charged jar, and the other on the knob of the jar so as to complete the circuit; the jar will be discharged, and the water will show the passage of the powers by being highly illuminated.

Some curious observations made on the electricity of silk by Mr. Symmer, appear in the Philosophical Transactions, which I shall briefly relate In cold dry weather, and chiefly when a north or north-east wind prevailed, he perceived that a new black and a new white silk stocking, made warm by the fire, and put on the leg, one over the other, became electrical on drawing them off together, without separating their parts; and he perceived they adhered together with such a force that it required sixty times their own weight to part them. Letting them remain some time in their united state, and then separating them in contrary directions, one by the heel and the other by the upper part, they retained the form of his leg for a considerable time. On presenting one of them to the other in their separated state, they approached each other; and when in contact, became pressed closely together. Trying the experiment with two pairs of silk stockings, one black and the other white, each stocking repelled its fellow, and attracted the contrary one. The dye of the black stocking imparts a property to it which the other does not possess, and thus, having different powers, they attract each other. The distension of the stocking when off the leg we may suppose arises from the repellent property of the same sort or power of electricity, as before described. Two black stockings will not exhibit electricity.

We observe that some substances are more perfect conductors of electricity than others ; these may be distinguished by being divided into two classes. The first or the most perfect conductors are metallic and animal substances ; but the latter are supposed to be such only while the vital principle continues to act in them. The second class are imperfect conductors, namely, water and all other fluids in an acid or oxide state. All these substances likewise differ from each other in the same class, as to their conducting power.

This leads me to say something of Galvanism, though it is a science I am not desirous of cultivating, and therefore have not fully considered its results; yet as it has caused much attention, and is frequently adverted to, I shall briefly relate the nature of a Galvanic circuit.

To form this circuit, three conducting substances are employed ; one of the first class and two of the second, or two of the first and one of the second. When the circuit consists of two of the first and one of the second, it is called of the first order ; when two of the second and one of the first, of the second order. It is necessary, in placing the single Galvanic combination, that the two bodies of the same class should touch each other at least in one point, while each touches the one of the other class. Many cruel experiments have been tried in this science, at which human nature shudders, to prove what has

been long confirmed—that the nerves of animals are their most
susceptible parts; for the fact is ascertained in the necessary
operation of amputation; when the surgical instrument pierces a
nerve, the most acute spasmodic sensation is communicated to
the patient.

In consequence of the intimate connection between the nerves
and muscles, a sensation imparted to one is immediately communicated to the other; so that, if an union be formed between a
nerve and its muscle, a similar sensation will arise in both. Though
the spasm may be imparted to an animal body by one piece of
metal only, yet it becomes more evident when a combination is
formed, as in the Galvanic circuit. An active Galvanic circuit is
usually formed of zink (a semi-metal), silver and water—a little acid
added to the water renders the combination more active. If these
metals and water be properly placed on each other, they will, on
applying a finger to the upper and under piece, produce evidence of
an active power. This effect is the greater, according to the greater
number of parts used. The manner of forming the pile is, by placing a piece of zink upon a piece of silver, and then a piece of
wetted paper or cloth, and so on alternately. I will state the methods
by which the effects of Galvanism may be rendered evident to the
sight and taste; as they are such as afford sufficient specimens
for our purpose.

Experiment.

To render the effects of Galvanism evident to the taste, place a
piece of silver on your tongue, and a piece of any other metal under
it; then, on forming a communication between the edges of the
pieces of metal, an irritation will be felt in your tongue, and an acid
taste be perceived.

Experiment.

Placing a piece of zink between the upper lip and the gum, on the nerve connected with the eye, and a piece of silver under the tongue, on striking the pieces of metal together, or touching both at the same time with another piece of metal, faint flashes of light will be seen in a dark place. Some persons suppose the effects of Galvanism to be the same as those of electricity, though their mode of operation differs considerably: that they have a near connection, is very evident.

That the nerves of animals are conductors of electricity is rendered certain by many experiments. The torpedo possesses in itself an extraordinary evidence of this fluid; for if the back and stomach of this fish be touched by both hands at the same time, placing one in opposition to the other, the electrical shock is felt, and the fish appears agitated: if instead of the hand a stick touch it, a numbness seizes the hand which holds the stick. That expert experimentalist Mr. Walker mentions this circumstance, and represents it as having an effect similar to that felt by the hand on touching a charged battery: the latter evidence he ascribes to the stick being a bad conductor of the fluid. The torpedo has not any apparent attractive or repulsive power. The effects of electricity perceived in this animal are also observed, in a superior degree, in the gymnotus or electrical eel, in which they are accompanied by a bright spark, visible in a dark place.

The electric fluid appears of different colours, according to the density of the medium through which it passes. When it is rare, it appears of a bluish colour; when more dense, purple; but when highly condensed, clear and white, like the light of the sun. The middle part of an electric spark often appears diluted, and of a red or violet colour; the ends are more vivid and white—probably from the resistance of the atmosphere condensing its surface. These effects are perceived by taking the electric spark through different substances.

Experiment.

Place an ivory ball on the prime conductor, and take a spark through it with a metal one, and the fluid will appear red. Do the same by a ball of boxwood, and the spark will be violet: the variation of the colour arises from the different texture of the substances through which it passes, which either increases or diminishes its density.

On spontaneous Electricity.

The apparent spontaneous evidence of the electric fluid in some substances, and in certain circumstances, is supposed to depend on the loose manner in which they contain it: for on the slightest increase of heat in such bodies, they exhibit the characteristics of electricity.

The tourmalin, a fossil stone found in the island of Ceylon, is one instance of this sort; for it exhibits signs of electricity by attraction on being heated: while heating, one side is found to possess a power contrary to the other: presenting a pith ball evinces this difference. The effects of electricity on the tourmalin appear only on the opposite parts of the stone, and not all over its surface: in this quality it accords with other electric substances. This electric when rubbed emits a strong spark; and the greater the degree of heat communicated to it, the stronger its electrical power.

Sulphur, melted in a glass vessel, shows spontaneous electricity: this also is supposed to arise from the action of fire on its particles.

Of atmospheric Electricity.

The identity of lightning and electricity was discovered and confirmed by Dr. Franklin, who has rendered this knowledge of the greatest importance to mankind, by his observations, inventions and communications. The mode pursued by this philosopher, of confirming his theory, was, by collecting the matter of lightning itself from the thunder clouds; for which purpose he used a common paper kite, flying it by a small flaxen string, to which a key was fastened, having a silken string to insulate his hand. Observing a thunder cloud, he raised the kite to its verge, or within the atmosphere of its electricity. When he observed the threads of the kite standing erect, as if on the conductor of an electrical machine, he ventured—and it was, indeed, a bold attempt!—to touch the key with his knuckle; then, to his great joy, a spark appeared.

After rain had fallen abundantly, and completely wetted the kite, he drew larger sparks from the key. This experiment was made by Dr. Franklin in 1752.

As the benevolent man rejoices in every opportunity of rendering his knowledge useful to his fellow-creatures, doubtless the doctor felt still more pleasure when he reflected that, in this discovery, he was, in certain cases, possessed of the means of disarming the thunder of its terrors; which service was most extensively useful in the country of North America, where he resided.

Caution respecting the Experiment with the electrical Kite.

The key which receives the electricity from the cloud should lie on the ground; but, notwithstanding this precaution, a cloud may be so highly charged with the electrical fluid, that, though the key carry it to the earth, its atmosphere may extend to a considerable distance, and act so powerfully as to destroy animal existence; therefore I do not consider this by any means a safe experiment; nor need it be attempted, as the identity of lightning and electricity is perfectly established by the preceding observations and experiments.

A hotter and drier state of the atmosphere contains a larger portion of electricity than a colder and damper one; accordingly we have more storms of thunder in summer than in winter, and in hot and dry than in wet and cold climates. From which observations we may suppose the atmosphere to be constantly more or less impregnated with electrical effluvium. The curious phenomenon called the water-spout, often seen in the seas within the torrid zone, may be thus accounted for by electricity. In those hot regions, when the lower clouds are saturated with the electric fluid, they may attract the water of the seas, raising it upwards; and the vapour of the cloud stretching also towards the water, these two may exchange powers or properties, and cause that concussion which sometimes is so great as to sink large vessels. To avoid this danger, the mariners endeavour to disperse the cloud by firing a gun through it. Wherever the accumulation of electricity resides, there must be an electrical atmosphere surrounding it. Hence when the two electrical properties of water and vapour nearly meet, the atmosphere or air is rendered more subtile; and consequently the surrounding air presses in to restore the equilibrium, so that the effects become concentrated.

The spontaneous discharge of electricity from a thunder cloud is sometimes seen on the masts of ships at sea, and on the points of steeples on land. Cæsar relates, that in a storm of thunder sparks issued from the soldiers' spears; which caused astonishment and consternation in his whole army.

Analogical Comparisons between the Effects perceived of Lightning and Electricity.

Lightning, or electricity, is conducted more readily by metals and fluids than by any other substances. The progress of lightning may be traced from the top of high steeples, where it first strikes; thence proceeding along the metal, till it loses that conducting power, when it explodes. Wood, brick and stone, being bad conductors of the electric fluid, are sometimes damaged by the lightning passing through them to the earth. But the most dreaded effect is that it sometimes exerts on animal nature; for should an animal body be near, or within the influence of, a flash of lightning, the fluid will leave the imperfect conducting materials above mentioned, and pass into the animal body, which will be injured, or perhaps deprived of life, by its power. It may be possible for a man holding a sword in his hand, with the point resting on the ground, to conduct the electric fluid, and remain unhurt; the steel being a better conductor than his body, and the point of the sword diminishing the power of lightning in passing through it.

Wood being a bad conductor of the electric fluid, we should avoid standing under a tree in a thunder storm; because the lightning will pass through our bodies in preference to the tree, they being better conductors of electricity. Though most liquids convey the electric matter, yet oil does not; for which reason nut-trees are particularly dangerous, as their oily sap wholly impedes the passage of the fluid. A vessel at sea is very liable to injury from a thunder storm, by its exposure, and some of its materials not readily affording a passage to the fluid.

Soldiers during a storm should unload their guns and set them upright on the ground, for then the lightning, striking the gun-barrel, is conducted by the metal into the earth.   In a thunder storm, the safest mode, supposing a person exposed to the elements, is to suffer his whole apparel to become entirely wet with the rain, which may prevent the bad effects of the lightning, should it strike near him.   In an apartment during a storm, it is safest to sit in a free current of air, and not against walls or chimneys; and all metallic substances should be discarded during a storm.

The facility with which metal conveys the electric fluid, produced Dr. Franklin's invention of rods to conduct the matter of lightning, and thereby to preserve houses and ships from destruction. He proposed this method in 1752, which was adopted in many places of North America, and particularly in those parts where lightning was most common.   He conceived that if a metallic rod were fixed above the upper part of an edifice, and continued downwards by other connecting portions of metal, and the other end were sunk some feet in the earth, the matter of lightning, striking on the top, would be conducted to the other extremity, without injury to the surrounding substances.   Perceiving also the silent effect of electricity, and the smallness of the stream from a pointed body, he terminated these rods by points.   After repeated trials of conducting rods in the provinces of Pennsylvania, Maryland and Virginia, he found them in general a preventive against the destructive effects of lightning; and where they did not absolutely effect that purpose, it was owing to the conductor not being within the circuit of the lightning's atmosphere.   Some persons cavil about the use of pointed conductors, though they admit of those with round tops, from an opinion that the point attracts the electric fluid; but many experiments contradict this idea, and one in particular presents itself in the circumstance of the magazine at Purfleet, which was furnished with a pointed conductor, and yet the lightning struck on an iron cramp at a distant corner of the building, in the side situated only a few yards below the conductor. This I think sufficiently evinces the fact, that the point does not considerably divert the lightning from its direction.

The conductors of lightning fixed to houses are doubtless uncertain security; yet as portions of metal always have place on the surface of large edifices, it becomes necessary to place a conductor communicating with these, in order to convey the electric matter to the ground. I conceive this to be a measure of consequence in all buildings; but, to render it a certain security, all the various portions of metal in an edifice ought to be connected in this manner, which is impossible; yet the larger portions of this material, which are situated in the roof of the building, may be so connected. It appears natural to suppose that pointed conductors are safer than knobbed ones, because they receive the fluid in a more gentle manner, and accordingly are not so liable to injure the edifice to which they are attached; for when the matter of lightning is received on large round surfaces, the violent percussion sometimes shakes the building, and thus destroys some of the advantages that might otherwise arise from its conducting power.

The different sensations produced in a person by the electricity received from a pointed and from a round surface, even from our weak accumulations of the electric fluid, naturally convey this idea; as does also the experiment of silently discharging a jar by means of points.

Experiment.

Charge a jar positively or vitreously, and expose the point at top

E E

instead of using the ball. Then, on discharging the jar, by presenting
a needle to the point at top, and a finger to the coating, the charge
will be drawn off silently. The popular experiment of the thunder-
house serves to exhibit the effects ascribed to conductors.

Let fig. 2, pl. 23, represent the end of a house, which in this experiment is a board of about half an inch thick, fixed perpendicularly on a horizontal plane. On this a glass pillar, A, is placed
within a few inches of the perpendicular board. A small part of
the upright board is cut out, about half its thickness, and a brass
wire, $b\,c$, laid into the piece of wood, 1, 2, which serves to form a
communication with one at e, and another at f; when this piece of
wood is put in loosely with the diagonal wire $b\,c$, not touching the
wires at $e\,f$, as represented fig. 2, it serves to show the effect of
lightning on a house having a broken conductor. When the piece
is laid in so as to unite with $e\,f$, as in fig. 3, pl. 23, it exhibits the
circumstance of a house having a continued conductor from the top
to the bottom. To the upper extremity of the glass pillar A is fixed
a sliding wire, which admits of the knob B being placed within a
quarter of an inch from the ball on the top of the house. We will
first fix the piece of wood, to represent a broken or imperfect conductor; then connecting the wire on A with that at the top of the
jar, and with another chain connecting the coating of the jar with
the brass hook at the bottom of the thunder-house communicating
with the conductor, we charge the jar. To discharge it, we slide the
ball gradually towards that at top of the house, not touching the
coating of the jar, and an explosion takes place, which forces out
the piece of wood $b\,c$. For a confirmation of the use of conductors we
alter the experiment, by placing the diagonal wire in union with the
wires, as in fig. 3; when, though the explosion will take place, the piece
will not be driven out. Lastly, unscrewing the ball from the conductor,

and leaving it perfect, as at fig. 3, on discharging the jar, the piece continues in, and no explosion is heard, because the points convey the fluid silently to the earth. Another inference, not usually made, may be deduced from these experiments. It has sometimes happened with me, that when the jar has been very highly charged, and the fluid been received on the knob, even though the conductor was perfect, the piece has fallen out by the violence of the percussion. To render that fact more satisfactory to me, I charged a jar even higher than in the former instance, and exposed the point, when the piece remained in. This convinced me that pointed conductors were certainly the best.

Having exhibited such a variety of experiments in electricity, and all that serve to establish generally-received opinions, I shall close this subject by a few more theoretical inferences drawn from them.

Throughout our investigations, we have perceived the identity of lightning, fire and electricity; such as their similar action on conducting substances, their appearances, and their powerful exertions to place themselves in equilibrium when their power is accumulated: from hence we conclude, that this fluid, under its different modifications of fire, lightning and electricity, pervades all nature.

We may suppose lightning to be the rapid motion of large quantities of the electric fluid, passing through the atmosphere to the earth, to restore the equilibrium it has lost by an accumulation of its power; that in passing through the atmosphere it causes a temporary vacuum, which the neighbouring clouds endeavouring to enter, the noise of thunder is produced. When the clouds containing the fluid are not very highly charged, the lightning appears, but the thunder is not heard.

It sometimes happens that persons killed by lightning exhibit no external marks of violence. This circumstance may be explained by the vital parts of the body being more delicate than those which compose its surface, consequently the former are more easily affected than the latter. Beccaria supposes, that the lightning does not really touch such persons, but deprives them of life by producing a vacuum in the atmosphere surrounding them; and, in support of this opinion, he says, that the lungs of such persons are found flaccid; whilst those on which external marks of lightning have been perceived, have had theirs inflated. I think it does not require this strained hypothesis to account for what our senses so readily explain. This leads me to say something of medical electricity.

Medical Electricity.

The agency of fire, in generating and continuing motion in the animal system, is a truth well known: this fire, under the form of electricity, shews itself in the human frame even without our attention in many familiar instances; one of which I am particularly sensible to in my own person. The atmosphere in very cold weather is nearly an electric; which when excited by friction, as in combing the hair, produces a crackling noise, and, in the dark, sparks are seen to issue from it; because the body possesses more of the fluid than the cold air surrounding the head; and an excitation takes place by the friction of the comb on the electric atmosphere.

The various modes of applying of electricity to palsied limbs, and for different affections of the human body, which have been attended with medicinal success, are too numerous and well-authenticated to require a particular statement here; those who wish for full information on this subject, I refer to the gentlemen who are most experienced in this application of electricity. I am so well convinced of its efficacy, that I recommend it as a branch of charity that is likely to be of extensive use to the lower class of society In my neighbourhood, Swift of Greenwich, a very expert and able electrician, under the direction of medical gentlemen, devotes his attention to this branch of the science in particular; and to the poor he very laudably affords this relief, at the moderate charge of only one guinea per annum; which contribution entitles a subscriber to have one patient continually on his list; by which means many persons deprived of free muscular motion have been restored to the use of their limbs, who otherwise would have been lost to the means of supporting themselves and their families. How great these advantages, and how easy the purchase! How truly gratifying to a liberal mind, to be thus instrumental to the benefit of society!

As to the species of complaints likely to be relieved by electricity, from the known effects of this fluid by experiments on various substances, we may suppose that its efficacy extends to all kinds of obstructions of motion and circulation.

From the preceding investigation of the nature of electricity by experiments, the only probable results are the following:—that the electric fluid, though in its natural state invisible, yet pervades all things, and is a principal agent in all the operations of nature: that when its equilibrium is disturbed it becomes visible, and endeavours to restore itself to its natural state by communication: that when two light bodies possess an equal power, or the same quality with each other, repulsion takes place;—may not this arise from the extension and equal action of their atmosphere ?—but when they are in contrary states, attraction takes place; produced by their natural affinity with each other.

As this lecture has exceeded my usual limits, I shall close it by briefly remarking, in reference to the different opinions on the nature and operations of electricity, that objections may be made to almost any theory on natural phenomena, however illuminated by human reason. That there have been such made to the best established facts, is evident in the case of the justly-celebrated Torricellian experiment of the weight of the atmosphere : in which invention was tortured to overturn reason by the most absurd hypotheses—Linus, satisfying the adherents of the fuga-vacuo, by asserting, that " the " mercury was supported from the top of the barometer tube by in- " visible threads !"

PLATE XXIV.

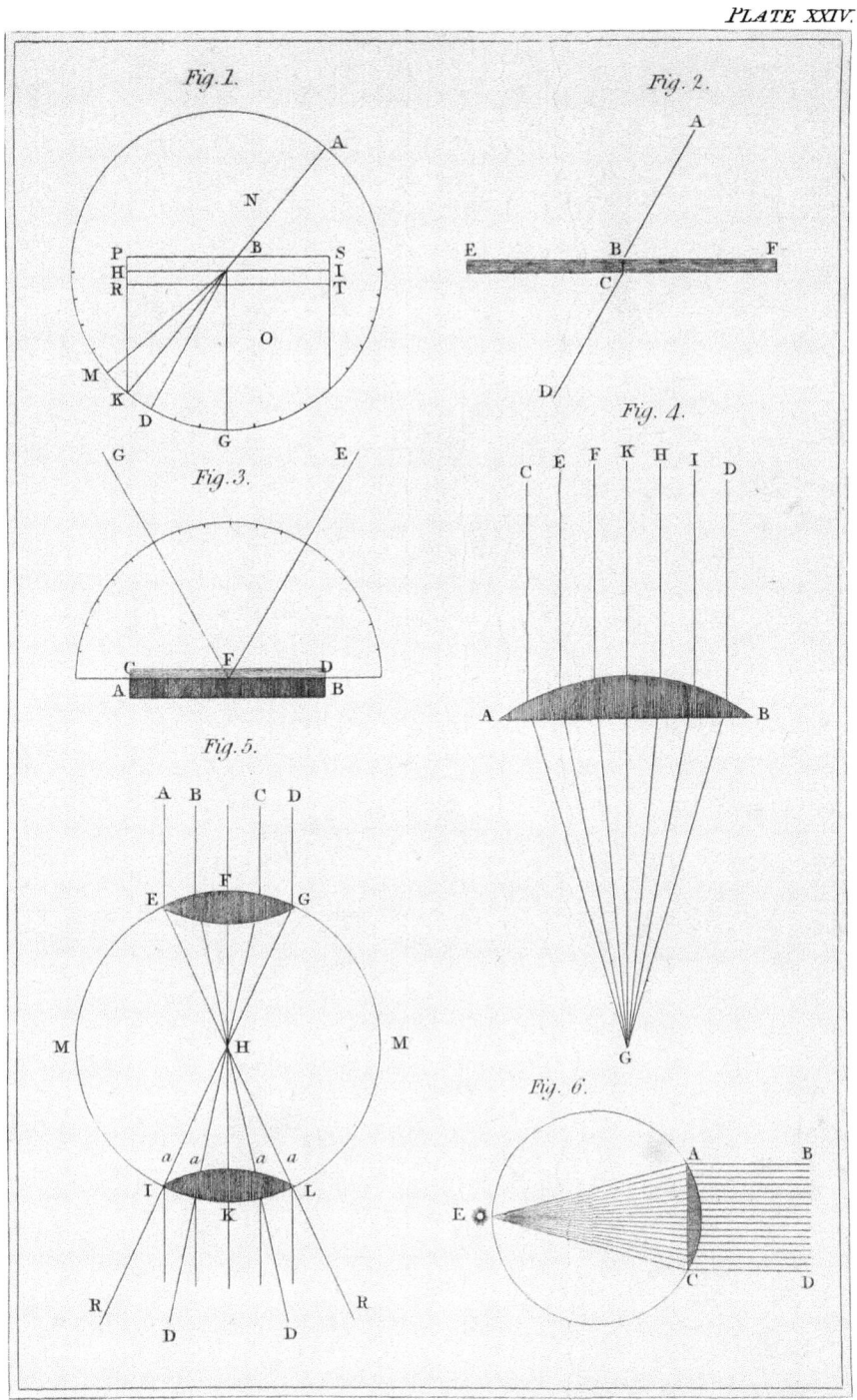

Fig. 1.

Fig. 2.

Fig. 3.

Fig. 4.

Fig. 5.

Fig. 6.

PLATE XXV.

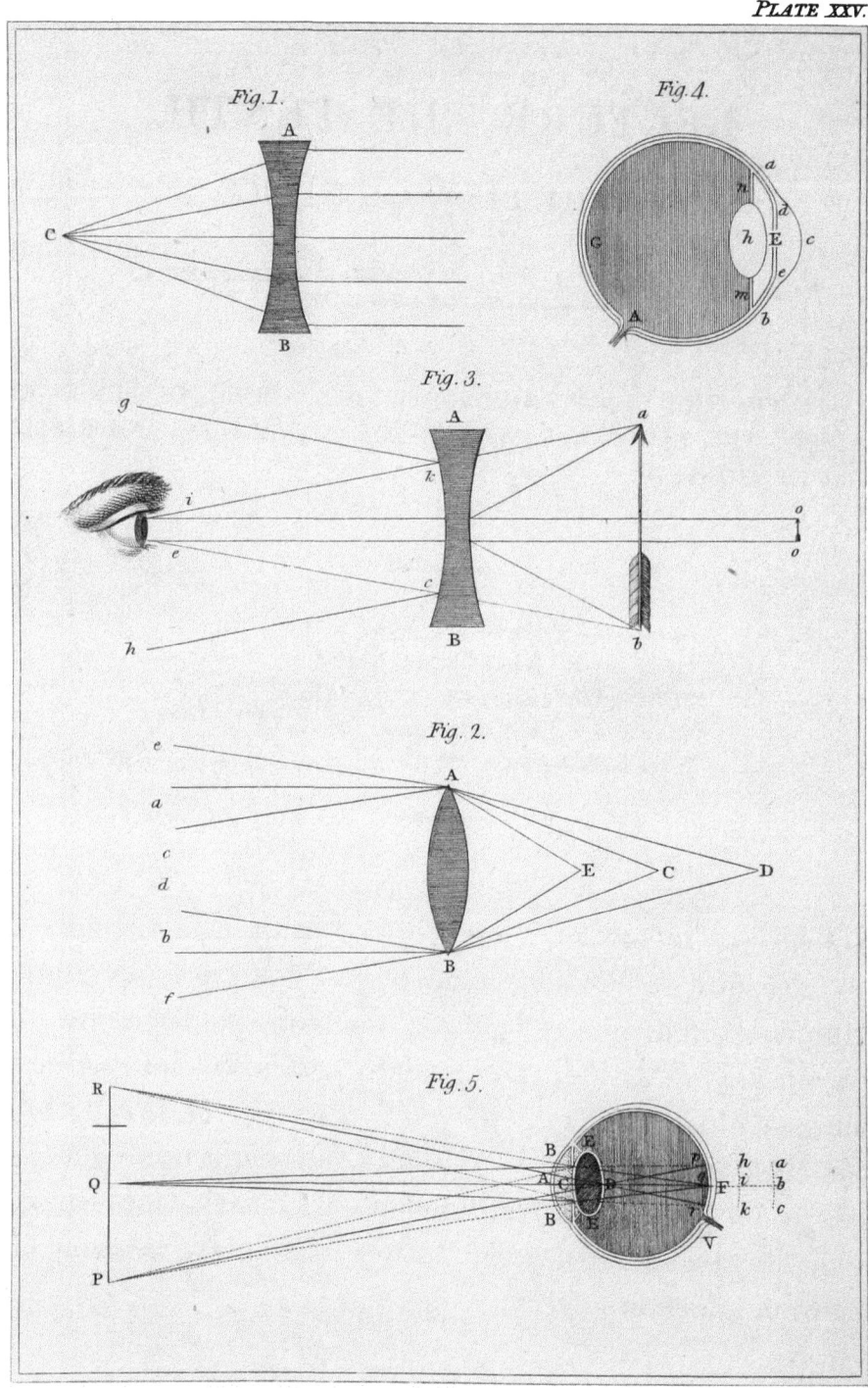

Fig. 1.

Fig. 4.

Fig. 3.

Fig. 2.

Fig. 5.

M.Bryan del.

H.Mutlow sculp.

LECTURE THE TENTH,

AND THE FIRST ON OPTICS.

THE AFFECTIONS OF LIGHT CONSIDERED.—THE NATURE OF LENSES EX-
PLAINED.—THE LAWS THAT GOVERN THE REFRACTIONS AND REFLEC-
TIONS OF LIGHT, &c.

———

———— Prime cheerer, light!
Of all material beings first and best!
Efflux divine! Nature's resplendent robe!
Without whose vesting beauty all were wrapt
In unessential gloom.   THOMSON.

———

NO subject in nature is embellished with higher graces than
optics. No gift of Providence more justly excites our gratitude
than the cheerful luminary of day: it charms all our senses, invigorates our animal nature, and expands, exalts and spiritualises our
reasoning faculties. It is the universal dispenser of sweets, to gra-
tify our smell and taste; and of beauty, to charm our sight. It is
nature's prime embellisher, and human life's chief solace—sparkling
in the diamond with resplendent lustre, and mildly beaming intelligence from affection's eye.

The term optics, properly, implies only the nature and properties
of vision; yet it is universally used in a more extended sense—as

including the nature and all the properties and affections of light. Agreeably to this use of the term, I shall begin this lecture on the subject by relating the present established and best-authenticated opinions on the nature of light, as communicated by Sir Isaac Newton, and confirmed by him so far as human reason, aided by experiments, can establish facts. He conceived light to be real matter, composed of particles differing in size from each other, and continually thrown off the surface of the sun.

The particles of light move in straight lines, and with amazing velocity, in which they would go on for ever, were they not stopped in their progress, or diverted from their rectilineal direction, by intervening substances.—The velocity of the rays of light amounts to the astonishing ratio of nearly 200,000 miles in one second of time!

The least portion of light that we can describe by a line is called a ray This is made up of innumerable particles following in a straight line, yet at a great distance from each other; for if they immediately succeeded, they would destroy instead of aiding the organ of sight. A ray of light is diverted from its rectilinear direction in two different ways—by the refractive or reflective property of the body it approaches, and sometimes by being absorbed or lost in the body When it passes obliquely through a medium of a different density from that in which it previously moved, it becomes distorted or turned aside, but it reassumes the rectilinear direction when it leaves the substance which transmits it.

When light strikes certain bodies with certain angles, it is reflected from them; but when it strikes these with greater angles, it is imbibed. The velocity of light has been established by different means, the most popular of which is the observations of the eclipses of Jupiter's satellites; and I shall briefly relate the application of these obscurations to this discovery, as conveying the most positive evidence of the fact.

It was first observed by Roemer, that the eclipses of Jupiter's satellites appeared either sooner or later than shown by the astronomical tables, accordingly as the earth was nearer to Jupiter, or further from that planet: in the former situation of the earth in its orbit, the eclipses appearing earlier, and in the latter, later, than the calculation previously made. After numerous observations, it is established, that when the earth is in that part of its orbit nearest to Jupiter, directly between that planet and the sun, an observed eclipse of a satellite happens $8\frac{1}{4}$ minutes earlier than the predicted time; and when the earth is in the opposite part of its orbit, the furthest from Jupiter, the eclipse happens $8\frac{1}{4}$ minutes later than foretold by the tables. Accordingly we infer that the motion of light, like that of all other bodies, is progressive, not instantaneous; that it traverses the whole earth's orbit, or a space of nearly 200 millions of miles, in $16\frac{1}{2}$ minutes, which is at the rate of 200,000 miles in a second of time! From this observation we also learn that the light of the sun is $8\frac{1}{4}$ minutes travelling from that luminary to our earth. The known velocity of light in passing from Jupiter to our earth, enables us to calculate the time that it is passing from any of the planets to our earth, and from the sun to each of the circulating worlds.

Sir Isaac Newton's theory and experiments on light and colours surpass all others; and are so clear, simple and well confirmed, that they must be credited by every discerning experimentalist and true philosopher: yet his assertion of the materiality of light has been objected to, on the supposition, that if the particles of light be continually projecting from the sun, that luminary must suffer a great diminution, and in time be entirely annihilated. This reasoning is not quite consonant with just ideas of the power, wisdom and benevolence of the great Creator, who endows all bodies with their requisite powers and capacities. Besides, it is a well-known fact, that from the small portion of the sun's light which is continually emitted, the whole substance of that body could not be exhausted in millions of ages.

Some persons suppose light to be merely the action of a subtile fluid, caused by the vibrations produced by luminous bodies, which thus convey the perception of objects to the eye. Whatever be the nature of light, it produces a sensation on our organs of sight, which renders evident to that sense the objects by which we are surrounded.

From the various affections of light, it is evidently a material substance; and as such I shall consider it in our investigations. I shall first generallise, in my communications, the universality and uniformity observable in the affections of light under certain circumstances; and afterwards treat of the nature and effects of light individually.

Light, like other bodies, is subjected to the laws of attraction; being sometimes diverted from its straight course by the interposition of substances, and at others retained in them. When it passes through a substance or medium of a different density from that by which it is transmitted, the attraction of the interposed medium causes it to deviate from its rectilineal direction; and this deviation is called refraction: when it is retained in the substance, it may be called absorption. Light is also capable of reflection, or turning back, in the medium in which it moves. When this happens from a hard polished surface, it is almost total; but when from softer and rougher surfaces, more partial. The particles of light are also capable of accumulation, dispersion and separation.

That light obeys the universal law of attraction, is evident by the various degrees of refrangibility produced on a ray in passing into mediums of different densities, according to the attractive powers of the mediums, and the different inclinations of the rays of light to them. These circumstances I shall endeavour to explain to you both by diagrams and experiments; and, preparatory to this discussion, shall acquaint you with Sir Isaac Newton's idea of the cause which produces the refraction and reflection of light.

He supposes that all bodies contain subtile effluvia; and that the action of this medium extends to some distance from their surfaces, occasioning the refraction of the rays of light in passing through certain bodies, and, by its reaction, causing them to be reflected from others.

This medium he imagines to be condensed in our atmosphere; and its density to be increased in the substances on the earth, and in proportion to the number of particles in the bodies it penetrates: that in its natural state it is rarer than any thing we can conceive; and occupies what is by us called space, or the ethereal regions.

" If light," says this great man, " were reflected by striking on
" the solid parts of bodies, their reflections would not be so regular
" as they are found to be; as, however polished the smoothest surface
" may appear to our sight, or feel to our touch, yet is it an assem-
" blage of inequalities; for in polishing glass with sand, it is not
" otherwise polished than by bringing its surface to a very fine grain,

" and from such a surface the small particles of light would be as
" much scattered as from the roughest." This reasoning, though we
cannot see the invisible agent that causes reflection, must convince
us of the probability of its existence.

That the rays of light proceed in straight lines, while moving in
the same medium, is evident, by the shadows cast by opaque bodies;
and that in falling obliquely from one medium on another of a
different density they are refracted, is proved by experiments.

A medium is any transparent body which suffers light to pass
through it; such as air, glass, water, &c. These mediums differing
in density, the light passing from one to another is refracted, or
turned from its straight course; and we may suppose this to be
effected by the attraction of the denser medium.

The rays of light falling perpendicularly on a denser medium
are not refracted; only those which fall obliquely on it.

A ray of light in passing out of a rarer into a denser medium,
is refracted towards the perpendicular of the latter; that is, towards
the axis of refraction: hence the angle of refraction is less than
the angle of incidence; whereas these two angles are equal when
the ray is not refracted. The physical cause of this may be, that
the attraction of the denser medium, acting perpendicularly to
the oblique direction of the incident ray, diverts it from its course.
But when a perpendicular ray falls on a denser medium, it suffers no
refraction, because the attracting power acts in the same direction
with the ray, or equally on both sides of it.

When a ray of light passes obliquely from a denser to a rarer medium, it is still refracted towards the perpendicular of the denser medium; again making the angle of incidence less than the angle of refraction.

I shall now, on Sir Isaac Newton's principles of attraction, attempt to explain the laws of refraction, referring to fig. 1, pl. 24. Conceiving the line II I to be the boundary of two mediums—suppose, for instance, air and water; the upper one, N, the rarer, air; the lower one, o, the denser, water—their attractions are in proportion to their densities; because the more compact or heavier medium possesses a greater quantity of attracting particles.

Suppose P s to be the distance to which the attracting power of the denser medium extends itself within the rarer. Let the line from A to B represent a ray of light falling obliquely on the boundary line H I, which divides the two mediums, or rather upon the line P s, where the denser medium begins to exert its power. The ray arriving at B, is turned out of its straight course by the superior attraction of the medium o, which draws it towards its centre or perpendicular. The ray is bent from its straight course continually all the way in every point of its passage between P s and H I, within which distance the attraction acts. Between these lines it describes an arch; but beyond H I, being out of the sphere of the attraction of the medium N, it proceeds uniformly in a right line from the point w to D; but is drawn from its incidental direction from A to K, and nearer to the perpendicular of the denser medium o.

Reversing the mediums, suppose N the denser and more attracting medium, and o the rarer one. The attraction now acting in lines perpendicular to the surface of the upper medium, the ray will be continually drawn upwards towards the denser medium, and from the perpendicular of the lower or rarer one; that is, from its straight course B K, towards B M: accordingly we perceive that when a ray of light passes out of a rarer into a denser medium, being refracted or drawn towards the perpendicular of the denser medium, the arch or angle of refraction D G is less than that of incidence K G.

When we suppose N to be the denser medium and o the rarer, and the ray to be passing out of the denser medium into the rarer; the refraction being, as usual, towards the denser medium, the angle of refraction M G is greater than the angle of incidence K G: as may be seen by the letters on the circle.

When a ray of light, as A B, fig. 2, falls obliquely on a plane glass, E F, though it is refracted in passing through it, yet it afterwards goes on in the same line of direction, as shewn by A B and C D; the former representing the incident, and the latter the refracted ray. The particles of light being of different sizes, are not all equally refrangible; the largest particles being the least, and the smallest the most, liable to be turned from their straight course: for all bodies require forces proportioned to their magnitude to give them a determined direction. Rays of light proceeding from a large body at a very great distance, as the sun, may be considered as parallel.

A luminous substance may be seen from all points to which a straight line can be drawn; therefore a luminous body, by some unknown power, returns rays in all directions; which effect is called reflection.

The direction of reflected light is subject to the same mechanical law as regulates the reflections of all bodies whatever, namely, the angle of reflection is equal to the angle of incidence.

Suppose A B, fig. 3, pl. 24, to be a mirror, and the lighter shaded part, C D, the reflecting medium on its surface, receiving an incident ray of light, E F. This will be reflected from F to G. The degrees cut on the semicircle by the two rays show their angles to be the same.

The particles of light being inconceivably small, and projected by the wise arrangement of Providence at a due distance from each other, they will pass through substances of the most compact nature, and in every possible direction, without confusion, or impediment to each other's motion and direction.

The independency of the particles of light in respect to their individual motion and direction, is made evident by placing a convex glass, an inch in diameter, in a window-shutter, and holding a white paper at a proper distance from it in a dark room, when a representation of all the external objects within a certain distance will pass through the glass, and the rays, crossing each other, will exhibit the objects reflected therefrom in an inverted position.

That the rays of light proceed only in straight lines from a luminous point in a uniform medium is evident by their never passing through bent tubes.

Rays radiating from a centre, diverge or spread out: hence it is, that heat and light diminish as the distance increases from the object that emits the rays; which diminution is in proportion to the squares of the distances. An object placed at twice a given distance from a luminous object, receives only one-fourth the heat or light felt by it when placed at the single distance.

A ray of light falling perpendicularly on the surface of a convex or concave lens is not refracted; but those rays which fall obliquely on it, are so. All parallel rays falling upon these surfaces are refracted, excepting the one which falls on the axis of the lens. Let A B, fig. 4, represent a plane convex surface, on which parallel rays, C, E, F, K, H, I, D, are falling. These rays, by the refractive power and form of the medium through which they pass, will be so refracted as to meet at G. If all the parallel rays from an object fall on a plano-convex lens, they will all meet in the same point, namely, at the extremity of the diameter of its convexity. A double convex lens is on both sides a portion of a sphere, like E F G and I K L, fig. 5.

If parallel rays, A, B, C, D, from an object infinitely distant, fall on a double convex lens, E F G, they will be so refracted as to meet in a point at H; then crossing each other at that point, the semi-diameter of the convexity of the lens, they will diverge to the contrary sides, and continue diverging, as from H to a a a a: but if another double convex lens of the same convexity be placed at the opposite side of the sphere of the lens' curvature, as at I K L, in the circle F M K M, the rays, after passing through the second lens, will leave it in a parallel direction; but the rays in crossing at the focus change sides. Hence the object will appear inverted. If the lens I K L were not interposed, the rays would go on, after crossing at H, in the direction H R, H D, H A, H R, and continue diverging more and more. Lenses which are formed of arcs of small circles have their foci nearer to them than those which are portions of larger ones. Convex lenses make parallel rays converge to a focus, increase the convergence of converging rays, and diminish the divergence of diverging ones. Hence they include many and great advantages: for, besides affording us the gratification of seeing objects at infinite distances, they also remedy the inconvenience that long-sighted persons would otherwise experience. As convex lenses cause the rays of light to converge more or less according to their degree of convexity, so do concave ones render them divergent in the same proportion. The properties of concave and convex lenses are directly contrary; for the concave lenses diverge the rays of light, rendering parallel rays divergent, diverging rays more divergent, and converging rays less convergent: hence they become useful to near-sighted persons.

A lens is a transparent substance of a different density from the surrounding medium, and terminating in two surfaces; either both spherical, or one spherical and the other plane, or both plane. A plane glass, as I have shown, will refract the rays of light that fall obliquely on it, but it does not collect them into a point. Convex lenses are called burning glasses, because the sun's rays condensed at the focus of a very large one, will consume almost any substance subjected to their collective power.

I will endeavour to explain the cause why convex lenses enlarge, and concave ones diminish, the angle under which objects are viewed through them.

Let the rays between A, B, C, D, fig. 6, pl. 24, be supposed to proceed from the sun. The rays from that luminary are neither divergent nor convergent, but are parallel at our atmosphere, and

G G

therefore fall in a parallel direction on a convex lens, as repre
sented in fig. 6, pl. 24.   Suppose A B a very large convex lens, on
which the sun's rays are falling; these will be so refracted as to
meet in a point on the other side of the lens, called its focus; which
spot is as far from the lens as is equal to the diameter of the sphere
of which its convexity is an arc, as at E.   Parallel rays passing
through a concave lens, are conveyed to a focus on the side turned
from the object that transmits them; as represented by the parallel rays from an object viewed by an eye at c, through the concave lens A B of fig. 1, pl. 25.   c is at the semidiameter of the lens'
concavity

We have already seen how parallel rays meet at the foci of these
glasses; but, to explain the familiar optical uses of them, I shall
show you by a diagram that oblique and parallel rays have not the
same focus.

Let A B, fig. 2, pl. 25, represent a double convex lens, and c the
focus of parallel rays, two of which are represented by a, b; these,
falling on the double convex lens A B, will come to a focus at c.
The diverging rays, c, d, from the same object, meet at D, and the
converging ones, e, f, at E.   The effects produced in objects when
viewed through a concave lens, may be understood from fig. 3, pl. 25.
Let $a b$ be the size of an object seen by the eye, $i e$. Then, supposing
the double concave lens to be interposed, the ray falling from the
top of the object at A, on the lens at A B, will be so refracted as to
diverge to g, and that from the bottom of the object at B will diverge
to h.   Thus all the rays between $e h$ and $g i$ will be lost to the eye,
$i e$, for only those rays from the object A B that fall on the glass be-

tween *k c* will fall on the pupil of the eye. Hence the object will be seen diminished as at *o o,* appearing also at that distance from the eye, at the supposed focus of parallel rays.

The extensive usefulness of lenses renders the subject of Optics truly interesting : every person is, some time or other, sensible of the good effects arising from this knowledge. Were the science of vision more generally considered, probably the imperfections arising from the decay of sight might be retarded, and some of their inconveniences entirely obviated.

To render our knowledge on this subject in some degree serviceable to mankind, I will state a few particulars worthy of attention. Long-sightedness arises from a depression of the lenses that compose the organ of sight, which is occasioned by a deficiency in the humours of the eye, and a rigidness of the muscles that regulate its movements. In order to retard and counteract as much as possible the inconveniences of this natural decay of sight, it is necessary, as we advance in years, to accustom ourselves to look at objects as near as we conveniently can, and to employ the eyes moderately in viewing attentively a variety of objects at different distances. To avoid the impressions of a very strong light, and the sudden transition from darkness to an opposite extreme, frequently open and shut the upper lid, in order to diffuse a fluid that is intended to lubricate the eye. This act is usually performed instinctively, without our attention; but as some persons, constantly occupied in contemplating very small objects, are apt to get a fixed position of the eye, I deem this caution necessary : I have myself experienced the inconvenience attending negligence in this particular, when painting small objects on ivory; for after sitting many hours, earnestly contemplating the performance, a painful sensation in my eyes has been produced. The muscles of the eye grow stronger by moderate exercise, and are enfeebled by disuse; and the coats of the eye are rendered flexible by use, and become rigid by inactivity.

Reading or writing by candle-light hurts the sight; but as our not doing so would be a great sacrifice of intellectual enjoyment, in order to prevent the bad effects of this almost indispensable habit, we should prevent a glare of light from falling on the paper we are contemplating. A green shade to the candles affords a cool reflection of light. Fires are very hurtful to the sight, both on account of the great glare to which they expose the eyes, and from the heat transmitted by them. But, with all our precaution, the organs of sight must necessarily become impaired as we advance in years; therefore it seems not improper, to mention the evidences by which the decay of sight may be perceived.

Seeing small objects indistinctly at the distance of twelve inches from the eye—requiring a greater degree of illumination than formerly to see objects clearly—frequent occasion to shut the eyes, to relieve them, after a little exercise in reading or writing—a mist obscuring objects viewed very near the eye—the letters of a book appearing double. When all these circumstances concur, or either of them individually is perceived, intelligent oculists and opticians recommend the use of glasses, but not at first such as are too great magnifiers. If the defect be merely want of illumination, a plane glass will answer the purpose: if to view the objects nearer, a small magnifying power is necessary. From these observations I am led to explain the nature of this curious, noble and comprehensive machinery; contrived by Infinite Wisdom! endowed with properties

which produce enjoyments nearly intellectual ; and with capacities
which are more extensively useful and indispensable, than almost any
other in the complicated mechanism of the human frame !

> Of infinite contrivance, matchless skill:
> Whether the site or figure we regard,
> Or distribution of the various parts,
> Perfective of the system, strokes appear
> Too exquisite for bungling chance to hit.
>
> BALLY.

PLATE XXVI.

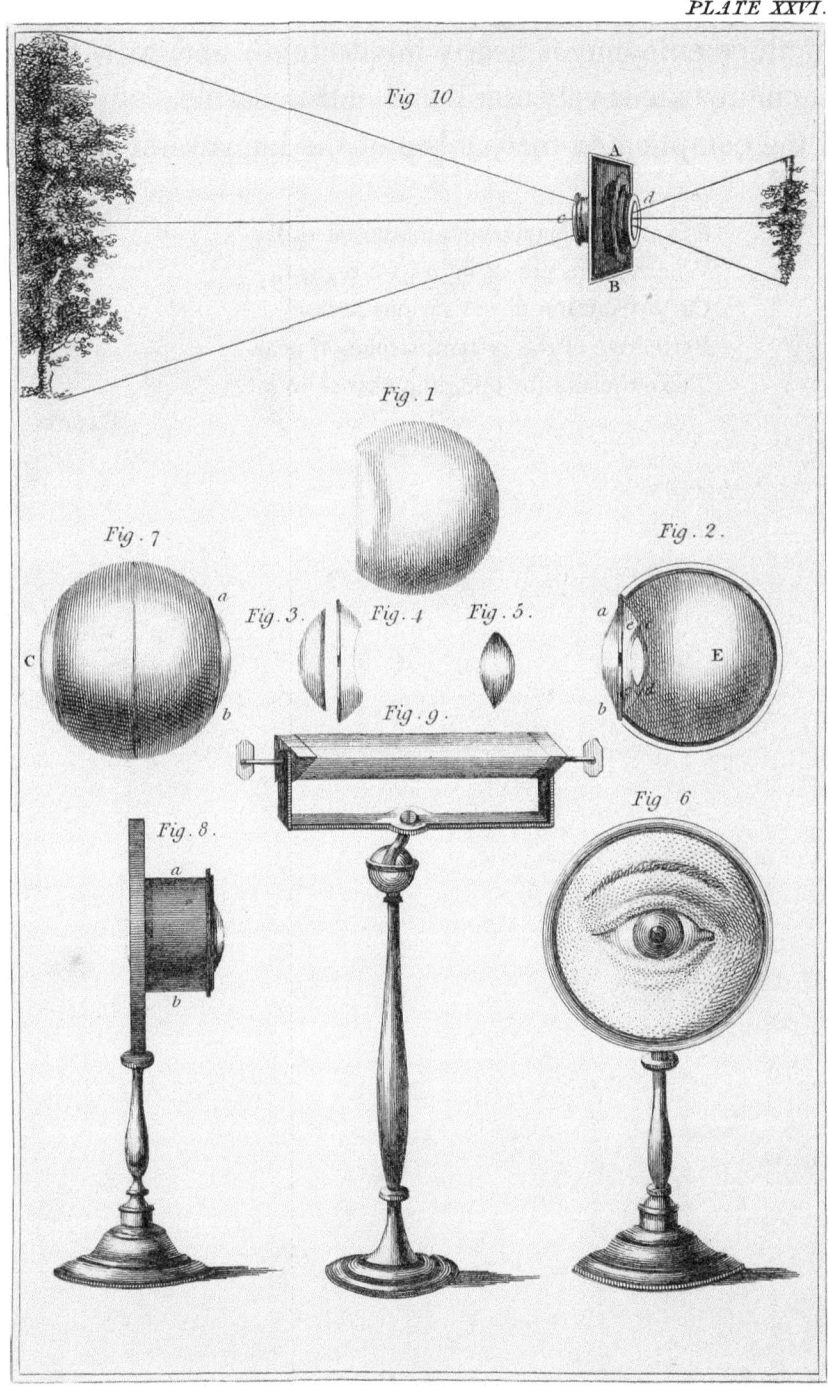

Fig. 10

Fig. 1

Fig. 7

Fig. 2

Fig. 3   Fig. 4   Fig. 5

Fig. 9

Fig. 8

Fig 6

T. Noble delin.

H. Mutlow sculp.

PLATE XXVII.

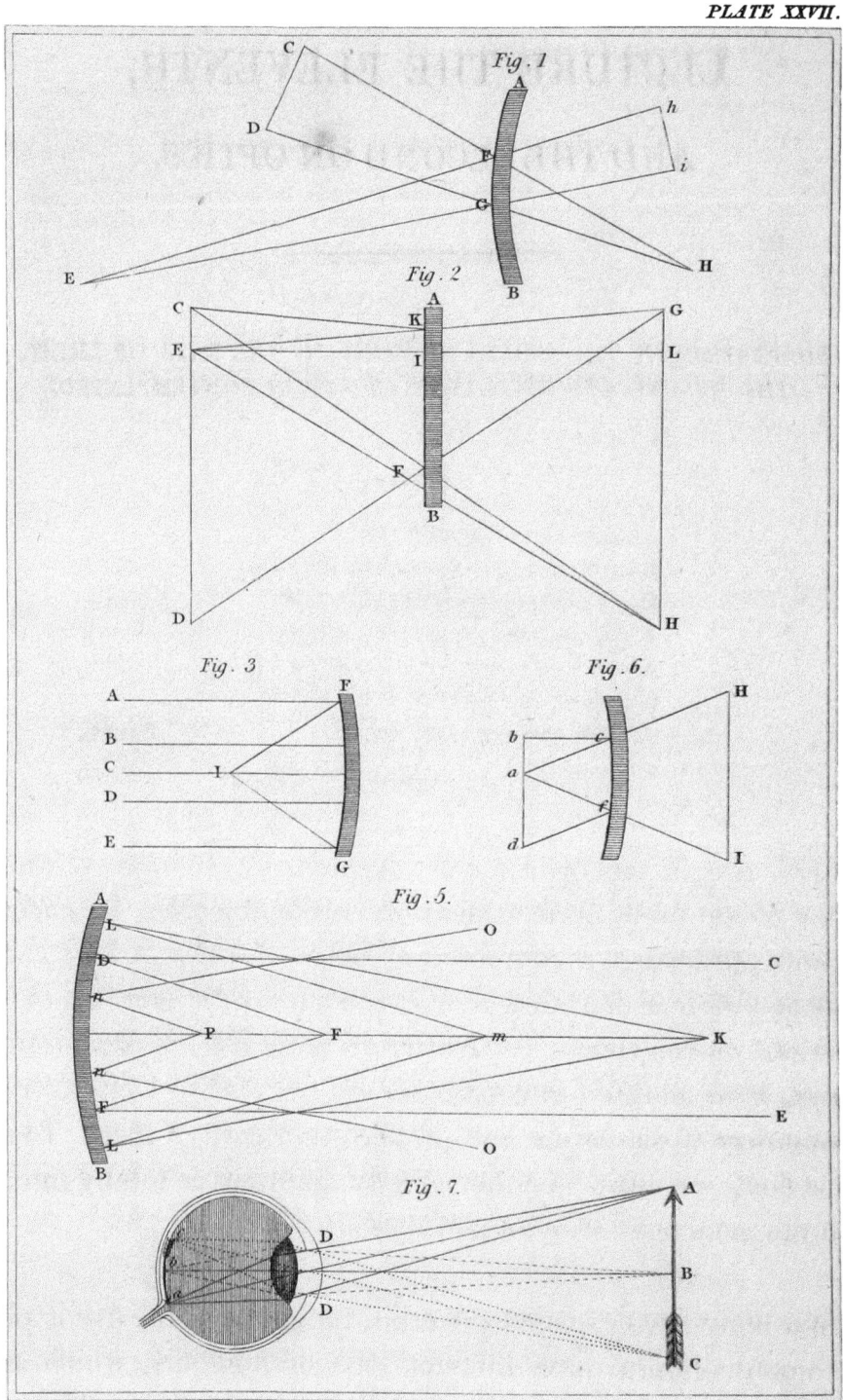

LECTURE THE ELEVENTH,

AND THE SECOND ON OPTICS.

THE ADAPTATION OF THE ORGAN OF SIGHT TO THE RAYS OF LIGHT, AND
THE NATURE AND AFFECTIONS OF VISION CONTEMPLATED.

———————— The visual orbs
Remark, how aptly station'd for their task;
Rais'd to th'imperial head's high citadel,
A wide extended prospect to command.
See the arch'd outworks of impending lids,
With hairs, as pallisadoes, fenc'd around,
To ward annoyance from without.   BALLY.

THE eye is the instrument invented by Infinite Wisdom to convey to the mind perfect ideas of visible objects. To effect this desirable purpose, it is composed of lenses of different foci, and mediums of different densities, and is supplied with muscles to effect its various movements. The matter of light, and the organization of the eye, were adapted to each other by the allwise Deity, on a due examination of the nature and properties of each of them. To doubt of this fact, we must be either grossly ignorant or totally insensible to all the evidences of sense and reason.

In a bony cavity, called the orbit, the globe of the eye is placed. This organ contains three different sorts of humours, which are enclosed in several distinct kinds of teguments or coats, in which blood-vessels, nerves and arteries are curiously interwoven. The orbit of the eye is supplied with a lubricating fluid, to facilitate the motion of the eye, and to preserve its delicate substance.

The eye-lids afford a perfect and secure asylum for the eyes when we sleep, and at all times diffuse a fluid over the eyes that keeps them constantly moist and clear; without which these delicate organs could not answer the purposes of vision: and when we are awake and unconscious of danger, the fringes that adorn them, by their exquisite sensibility, like the feeling organs of insects, cause the eyes to close against the approach of external annoyance.

These lids join at their extremities; and, to prevent their falling into wrinkles when open, each edge is stiffened with a cartilaginous arch, and is ornamented by a border of hair; which softens the contour of the eye-lids, and in the manner before observed, by its extreme susceptibility, protects the eye from straggling motes, and also moderates the light in approaching the retina.

In the upper part of the orbit of the eye a gland is placed, to convey the fluid that lubricates and polishes the eye. The corner of the eye next the nose is provided with a caruncle to carry off the superfluous moisture from the eye-lids. The eye is furnished with six muscles; these spread their tendons over the eye to regulate its motions in every direction, excepting in an oblique one, towards the nose, which is effected by a peculiar auxiliary. The side of the eye next to the nose not allowing room for a muscle, a small bone is placed there, having a perforation serving as a pulley for the tendon of a muscle to pass through, by which an oblique direction of the eye is obtained. What a curious contrivance and complete compensation! The eyes have a parallel or consentaneous motion, in which they always agree: this to human reason is extraordinary, as the organs of the two eyes are totally distinct. The purpose effected, as is supposed, by this union of action and direction, is seeing things single which are viewed double. According to Sir Isaac Newton, the species of objects seen with both eyes may unite where the optic nerves meet, before they come into the brain; the fibres in the right side of both nerves uniting there, and after union going thence into the brain. The nerve which is on the right side of the head, and the fibres on the left side of both nerves, unite in the same place; and, after union, go into the brain in the nerve which is on the left side of the head: these two nerves thus meeting in the brain in such a manner that their fibres make but one entire species or picture, half of which is on the right, and the other half on the left side of the sensorium. The optic nerves of all animals that look the same way with both eyes, as men, horses, dogs, &c. meet before they come into the brain; but the optic nerves of such animals as look in a contrary direction with each eye, as fishes, do not meet before they go into the brain. This conjecture is rational, and may therefore be admitted.

Undoubtedly vision, or the appearance of objects, is occasioned by the pictures on the retina; but it is the mind that perceives and judges. The eye is only an instrument by which the idea is conveyed to the mind; but the operations of the mind and its subtilty of perception we are not able to comprehend.

Considering the eye as it really is, merely an instrument, we need not enquire why the pictures of objects painted in a reverse position in it appear upright to our imagination. This difficulty anatomists are unable to solve; for it is impossible to afford a rational solution of a circumstance independent of the organization of the animal body. All our senses are aided by the mechanism of the organs created for their use, but their impressions must after all be refered to the spirit—the understanding; and therefore are not definable by human comprehension.

———————————— To attain
The height and depth of God's eternal ways,
All human thoughts come short.

MILTON.

The globe of the eye consists of three coats of different textures, containing the same number of pellucid mediums; these are so formed, that the rays proceeding from luminous objects, and received at the fore part of the eye, after entering the pupil, are brought to a focus on the back part of it, called the retina.

The sclerotica, the exterior of the three coats, is a hard substance, of a whitish colour, resembling parchment; represented in fig. 4, pl. 25, by the outermost circle of the three: its hinder part is thick and opaque; thence it becomes gradually thinner and thinner as it approaches the part in front of the eye, where the white terminates. The other part of this tegument is thin and transparent, and, projecting a little, forms the segment of a smaller sphere, like *a b c* in fig. 4; this part of the sclerotica is called the cornea, from its horny texture. It is perfectly transparent; which quality is necessary for the free admission of light. The sclerotica is composed of several layers, furnished with pellucid vessels, and replenished with a fluid medium.

The choroides, or second coat of the eye, is represented by the second circle drawn round fig. 4. It is of a soft and tender nature, and composed of innumerable little vessels. It adheres to the sclerotica. Its colour is brown exteriorly, and within side it is almost black. This tegument, like the sclerotica, is distinguished by two names; the fore part, situated between *d e,* is called the uvea, and the hind part the choroides. The uvea commences where the cornea begins. It is attached to the sclerotica by a narrow circular rim, from which part the choroides divides from the sclerotica, or changes its direction, turning towards the axis of the eye, forming the aperture called the pupil. The uvea commences where the choroides divides from the sclerotica; from this part to where the choroides forms the pupil, is called the iris. The colour of the iris is produced by the dark colour of the uvea, combined with the reflections of the rays of light that fall on that membrane.

The pupil of the eye, E, has no determinate size; for the choroides expands or contracts itself, in order to accommodate the size of the pupil of the eye to stronger or weaker impressions of the particles of light; as thus: when the light is too intense, this membrane expands itself and contracts the pupil, to prevent the admission of too great a quantity of light, which would injure the sight; but when the light is weaker, the membrane contracts itself and expands the pupil, and thereby a greater quantity of the rays of light falls upon the retina. These varieties regulate the action of the light on the retina, rendering it in both cases duly active.

The third membrane of the eye, or the retina, is spread like a net over the bottom of it, and lines the inside of the choroides; it also covers the concave surface of the vitreous humour, terminating where the choroides turns inwards. The inner circle in fig. 4, pl. 25, and the net-work included in that circle, represent this membrane, on which, at the back of the eye, the images of objects are painted; it being a continuation of the optic nerve, A. These coats contain the humours of the eye: one humour is nearly a solid substance, another perfectly soft, and the other fluid. The humours are of such forms and degrees of transparency as are best adapted for transmitting the rays of light, and placing them in positions favourable to distinct vision. These humours are all clear like pure water, possessing no essential colouring particles.

The most fluid humour of the eye is called aqueous, from its resemblance to water. It fills the space E *d c e*, between the cornea and the pupil; and also that between the pupil E and the crystaline humour *h*. Its quantity is so abundant, that it protrudes the fore part of the eye into the segment of a smaller sphere than the whole globe of the eye. Whence this humour is supplied is not known, but its source seems unfailing; for when the coat that contains it is wounded, and the humour flows out, by keeping the eye closed a proper time for the wound to heal, this fluid is perfectly recruited.

The second humour is called crystaline, being the most transparent of them all. It is less in quantity than the aqueous, and of a more dense nature. Its form is doubly convex, like *h*, fig. 4, pl. 25: its two sides are of different convexities; the more convex part is received into an equal concavity in the vitreous humour. The crystaline is contained in a case, thick and elastic in the part towards the front, and thin and soft in that towards the back of the eye. This lens is supported in its place by muscles, represented in fig. 5, by *n n*, which divide the globe of the eye into two unequal portions; the smaller and foremost containing the aqueous, and the larger and posterior the vitreous humour. The crystaline humour has no visible communication with its case, for when the latter is opened the humour slips out. In old age, the crystaline becoming discoloured, all objects appear less bright, and are tinged with yellow. The latter degeneracy coming on gradually, old people are not sensible of its effect.

The vitreous or third humour of the eye is like glass; it is neither so dense as the crystaline nor so fluid as the aqueous humour. It fills all the space between the sclerotica from the insertion of the optic nerve to the crystaline lens, and is included in the net-work, fig. 4. The optic nerve, A, passes from the seat of the brain through a small hole in the bottom of the orbit of the eye. It enters this orbit in a form nearly globular, and is inserted in the globe of the eye, but not exactly in the centre: from its appearance, it is supposed to be a continuation of the retina, being the part of that membrane which conveys to the brain the impressions of objects painted in the eye by the different refraction of the rays of light by the coats and humours of the eye, which cause distinct images of exterior objects to be painted on the retina. The humours of the eye are admirably contrived for effecting these purposes, for they cause all the rays of light entering the pupil from a luminous object to unite together on the retina, by which means they make a stronger impression than in their simple state. The retina is placed at such a distance from the refracting substances, that each pencil of rays received on it meets in distinct foci; and thus the images appear conspicuously, and are correct in their proportions. The degrees or powers of refraction of the humours of the eye have not been determined for all states of it; those of the vitreous and aqueous humours, in their perfect state, are supposed to be the same as common water, and the power of the crystaline a little more refractive. I will endeavour to convince you, by a diagram, that the different forms of the lenses which compose the globe of the eye are all essential to distinct vision.

Let P Q R, fig. 5, pl. 25, be a luminous object. The pencils of light B P, B Q, B R, from the points P, Q, R, falling upon the cornea, B A B, are so refracted that they would belong to foci beyond the eye, at *a b c;* but the surface of the crystaline humour, c, increasing the degree of refraction, would make them meet in foci nearer to the eye, yet exterior to it, at *h, i, k:* lastly, the refraction being still augmented by the pencils of light passing out of the crystaline humour into the vitreous, they are brought to their proper foci on the retina at *p q r*. But the manner in which distinct vision is accomplished by the different lenses and mediums of the eye requires a further investigation; with which I shall proceed, after showing you the exact configuration of the lenses that compose this curious organ by a dissection of the artificial eye, represented fig. 6, plate 26, and describing its individual parts by figs. 1, 2, 3, 4, 5 : these exhibit the exact forms of the different humours and lenses of the eye. Fig. 7, pl. 26, is a side view of the artificial ball of the eye, taken from the socket, having the fore-part, *a b,* projecting into the portion of a smaller sphere than the whole ball of the eye. c represents the opposite or back part of the eye, where a piece of the metal is cut off in this apparatus, to show the images on the part that represents the retina; for when the fore part, *a b,* of the eye is turned towards an object, the image of it appears on the retina at the back part, c, but inverted; the cause of which I shall explain when treating of vision.

Fig. 2, pl. 26, exhibits the humours of the eye properly placed, and the three circles encompassing them the three coats of the eye.

Fig. 3, represents the first chamber of the aqueous humour of the eye, in form like a plano-convex lens, as shewn by *a b,* in fig. 2 ; this lies over the part of the choroides that turns inwards. Fig. 4, exhibits the form of the second chamber of the aqueous humour, which is plano-concave, and is between the iris and crystaline lens, as seen by *e e*, between *a b* and *c d*, fig. 2.

Fig. 5, pl. 26, shows the form of the crystaline lens, a double convex of unequal convexities, situated as seen by *i d* and E, fig. 2 ; the more convex side being received into a correspondent concavity in the vitreous humour, E, fig. 2.

Fig. 1, pl. 26, is the form under which the vitreous humour is contained ; and the three humours represented by the figures drawn express their relative proportions to each other. Fig. 8, is a side view of the artificial eye, placed in its socket, *a b*, in which it moves

We have sufficiently contemplated the curious mechanism of the eye to comprehend its optical effects. To investigate minutely the variety of substances that compose this wonderful machinery, and duly appreciate the actions of all its parts, exceeds all bounds of human performance and inspection.

I shall now proceed with the science of optics, which embraces so many sublime phenomena, that it may truly rank as the most noble and pleasing contemplation in the whole charming circle of divine evidences, displayed in the subjects of natural philosophy Acknowledging the materiality of the rays of light on the authority of Sir Isaac Newton, and the evidence of our senses, we may now consider the other parts of the Newtonian theory.

Fig. 10, pl. 26, represents the scioptric ball ; it consists of a ball and socket, similar to the natural eye, by which contrivance it may be turned in various directions, like the natural organ of vision. Two lenses are placed in the hollow ball A B, one at *c*, the other at *d*.

This instrument being placed in an aperture in a window-shutter, and the room darkened; on holding a white paper at a proper distance from it, a distinct picture of the external objects opposite the ball will appear, but in a reverse and an inverted order. The rays of light falling from objects on the side of the ball presented towards them, cross each other within the ball, and therefore the objects are inverted and reversed. All the rays of light from a number of distant objects making but a small angle at the ball, pass through it; and though these rays pass through it in all directions, some parallel and others oblique, yet no confusion arises. This is a convincing proof of the separation of each particle of light, though, on account of the velocity of light, these particles appear to follow close to each other.

To remedy the unnatural appearances of objects when painted in an inverted order, a mirror has been added to this apparatus. I shall explain the nature of this improved camera when describing optical instruments; previous to which, it is necessary to form correct ideas of the laws that govern the reflections of light, properly called ca-toptrics. The first and most important fact to be understood in catoptrics, and which I have already explained, is this—that the angle of reflection is always equal to the angle of incidence. A person viewing himself in a plane mirror, appears of the same size in it, and at a like distance from its surface, as he would if viewed by another person at the same distance from himself that he is placed from the glass. A mirror of only half the length and size of a person is required to reflect his whole figure: these facts are familiar, but the laws that produce them require an explanation in these lectures. Rays of light have the same inclination to each other, and to the reflecting surface, after reflection that they had before it, the angle of reflection being always the same as the angle of incidence. As the rays from a mirror are reflected in a more divergent state, the object appears of the same size and distance from the reflecting surface, as if the rays proceeded to the same distance behind the glass that the person stands before it, as may be understood from A B, fig. 2, pl. 27, which represents a mirror. The line C D is the height of a person viewing himself in the glass A B. The eye of that person, at E, traverses the mirror from A to B, and he sees his whole person reflected from it, because the exterior rays from C D fall on the mirror A B, and therefore all the rays from his person included within that distance must be seen by reflection. The rays from C fall on the mirror at F, and those from D meet in the same point.

Suppose E, fig. 1, pl. 27, an eye looking at an object, C D. The rays from C D falling on a convex glass, A B, in an oblique direction, the reflected image will appear to the eye in the direction from F and G to E, because we always see objects in the direction in which they immediately strike the eye, and the object will appear behind the mirror, as at *h i:* had the exterior rays, C D, which fall on the convex mirror A B, and are reflected to the eye at E, proceeded without being turned back, they would have met in a point at H; but being reflected from the mirror, more diverging from the perpendicular of it, they make a smaller angle with the eye at the focus of parallel rays, *h i,* and hence the object appears diminished.

Let A, B, C, D, E, fig. 3, pl. 27, be supposed parallel rays falling on the surface of a concave mirror, F G; these are so reflected as to meet in a point, I, equal to half the radius of the mirror's concavity

I I

This is called the focus of parallel rays, and the real focus of the mirror.

The foci of parallel, convergent and divergent rays, are different; and I shall endeavour to explain these differences in regard to reflected rays from concave mirrors.

Let A B, fig. 5, pl. 27, represent a concave mirror; C D and E F, two parallel rays falling on it at D, F. On striking the reflecting surface these are reflected converging, and meet in a point at F, the semi-radius of the mirror's concavity, called its true focus. Let K L, K L, exhibit two diverging rays proceeding from the centre of an object; these also are reflected converging, though not so much so as the parallel rays; therefore they meet in a point more distant from the mirror than the focus of parallel rays, as at M. Lastly, let N O, N O, be two converging rays proceeding from the extremities of an object, and falling at *n n;* these being convergent, before they fall on the reflecting surface, will meet in a point at P, between the mirror and focus of parallel rays.

Understanding from the diagram that the focus of convergent rays is nearer to the surface, and that of divergent rays further from that of a concave mirror than its real focus; also that the object is seen distinctly only when between the mirror and its principal focus, which is at one-fourth the diameter of its concavity; we shall comprehend why the object appears enlarged within that focus, and inverted beyond it.

Suppose *a,* fig. 6, the eye of a person viewing his face in a concave mirror, and *b* the ray from the top of his head proceeding to c;

this will be reflected to his eye at A, in the direction $c\,a$; and that from the bottom of his chin at d in the direction $a\,f$: he seeing the object in the direction in which these rays strike the eye, and thus he appears to be at H I, and magnified in the proportion that II I bears to $b\,d$, by seeing himself under a larger angle, as before explained by fig. 5. This effect takes place only at the focus of parallel rays: beyond which distance the object appears diminished and inverted, by the rays crossing each other at that point, having their foci at a greater distance from the mirror than they are placed, and being seen under a smaller angle; this may be understood by fig. 5; and the figure appears in the air between the object and the mirror. This latter effect may be exhibited by any object presented at a certain distance from a concave mirror, and is a curious spectacle; which, when disguised, surprises those not previously acquainted with the cause that produces the effect perceived. To persons fond of optical delusions, I recommend the very entertaining and instructive Philosophical Recreations of Dr. Hutton, and other ingenious publications.

Having explained all that is necessary respecting plane, concave and convex surfaces, to enable you to comprehend the causes of their effects, I shall proceed in the important investigation of the nature and effects of vision.

Let A B C, in fig. 7, pl. 27, be an object from which three pencils of rays from A, B, C, fall on the cornea of the eye D; these will be so refracted by the coats and humours of the eye, as to converge and meet at a, b, c, on the retina, on which a distinct but diminished figure of the object will be painted, the retina being situated at the back part of the eye at a proper distance to receive it. That vision is thus produced may be observed in the eye of a bullock taken from the head immediately after the animal is killed: cutting off the thin coats at the back of the eye, then placing the organ in a dark chamber facing a strong light, and holding a piece of thin white paper at the back of the eye, the figure opposite the eye will be seen on it, in an inverted position.

The object *a b c*, in fig. 7, forms an angle at the back of the eye proportionate to the convergence of the rays produced on passing through its different coats and humours, which convergence is necessary to distinct vision.

The images of objects are distinct only at the axis of the eye; hence we cannot see the whole extent of a large surface distinctly at one time, though some of the rays flowing from great distances and remote parts fall on the retina; but to consider each division correctly, we must shift the position of the eye. Accordingly, to view objects with attention, both eyes must be turned the same way. The two images of objects unite in the brain, so as to form but one impression of the object viewed. If the axes of both eyes did not unite in one point, objects would be seen double; therefore the seeing objects single with both eyes, is in a degree dependent on the mechanism of those organs, though the sensation communicated by their impressions to the brain is the cause of the idea being conveyed to the understanding.

Sir Isaac Newton supposed that the natural rays of light, after passing through the different coats and humours of the eye, by actual strokes form the impressions of objects on the retina, and in the following manner. An impression being made on the substance of the retina by the active particles of light, is conveyed, by the reaction of that substance, to the sensorium; and occasions, in a manner unknown to us, the perception of external objects.

This subject is wrapt in too much obscurity for very distinct ideas to be formed on it; but of this we are assured, that the images on the retina communicate the impression to the mind: here we must stop, for in the contrivance of vision, as well as in all the arrangements of powers and capacities in the works of God, there is a boundary to our researches never yet exceeded by man, and such a one as human reason never can exceed, for the superior energies of the Divine mind are inscrutable. Such parts of the mechanism of our organs as God has been pleased to render evident to our senses may be appreciated by us as to their uses and capacities; but no further can we go.

How admirably is the eye accommodated to the various circumstances to which it is subjected! To view objects both in a very weak and a very strong light, peculiar endowments of power are absolutely indispensable: as without sufficient illumination, vision would be feeble and indistinct; and too great a quantity of light, by producing too strong impressions on the retina, might cause blindness, by destroying its delicate texture. To obviate the former inconvenience and the latter misfortune, the Almighty has endowed the eye with the power of enlarging or contracting the pupil, that the just proportion of light necessary for distinct vision, and no more, may be suffered to approach the retina. It was also essential to the human eye that the pupil should retain its circular form under all its variety of dimensions. It is said this is effected by a particular position of the regulating movements, and by which mode of action alone it could retain that form. The pupil of the eye is formed by the choroides: this, when the light is too vivid, instinctively contracts the pupil; but when weaker, enlarges it.

Another striking evidence of the wisdom and benevolence of

the Deity presents itself in the contrivance and regulation of motion in the various lenses that compose the eye, to enable this organ to view objects distinctly at different distances.   By considering the nature of lenses, and the construction of optical instruments, you will perceive that this could not have been effected without a proper adjustment of the different lenses which compose the eye. How this advantage is produced is not easily discovered: yet as there is an evident necessity for these adjustments, conjectures have been formed which appear reasonable; and which, I believe, have received confirmation from the researches of anatomists.   It is said that there are muscles so placed in the eye as to effect these changes; the cornea and crystaline being thus rendered capable of projecting and flattening themselves: by the former operation the axis of the eye is lengthened, and by the latter it is shortened; which effects are varied thus: If the object we are viewing be near, the cornea and crystaline become more prominent, and consequently the axis of the eye is lengthened; when the object is more distant, the cornea and crystaline are rendered flatter, and the axis is, in consequence, shortened.   From these observations it is evident, that when the humours and coats of the eye become flatter, as in old age, we should use convex glasses, and always endeavour to get near the objects at which we are looking, rather than accustom the eye to view things at an unusual distance; agreeably to these observations we regulate our optical instruments as circumstances require, to produce distinct images; for instance, when an object is near, we lengthen the tube, if distant, we shorten it*.

---

* For the comparative anatomy of the eyes of men and the inferior animals—each peculiarly adapted to the various necessities of the different species—I refer my readers to the excellent Dr. Paley's Natural Theology, which is replete with evidences of what the inspection of the human eye so clearly determines—the existence of a wise, benevolent and superintending Providence.

In adverting to the diagram, fig. 7, I have further to remark, that when a small object falls on that part of the retina which forms the commencement of the optic nerve at E, its figure is not seen by us. Place a very small bright or opaque object so as to be very evident to your sight: then covering one eye, move the other nearer to your nose, which is the direction of the optic nerve, and you will not see the object. The object may in the same way be lost to both eyes by a certain direction of them, which I shall not communicate, as the performance might produce bad effects on the organs of sight. How astonishing are the powers of the eye! How wisely adapted and duly balanced! Yet we cannot see half the wonders of its structure; but we see enough to compel us to acknowledge the power, wisdom and goodness of the Deity—

> Whose matchless skill the needful eye did form,
> And with intelligence invest.   Then its
> Station fix'd; its tender form defended;
> With grace and harmony its front adorn'd!
> Thus form'd, appointed, defended and endow'd,
> The curious organ, as instinctively,
> Performs its various uses.   Oh, divine
> Contrivance! exquisite effects! surpassing
> Human skill—as much as God excelleth
> Man!

PLATE XXVIII.

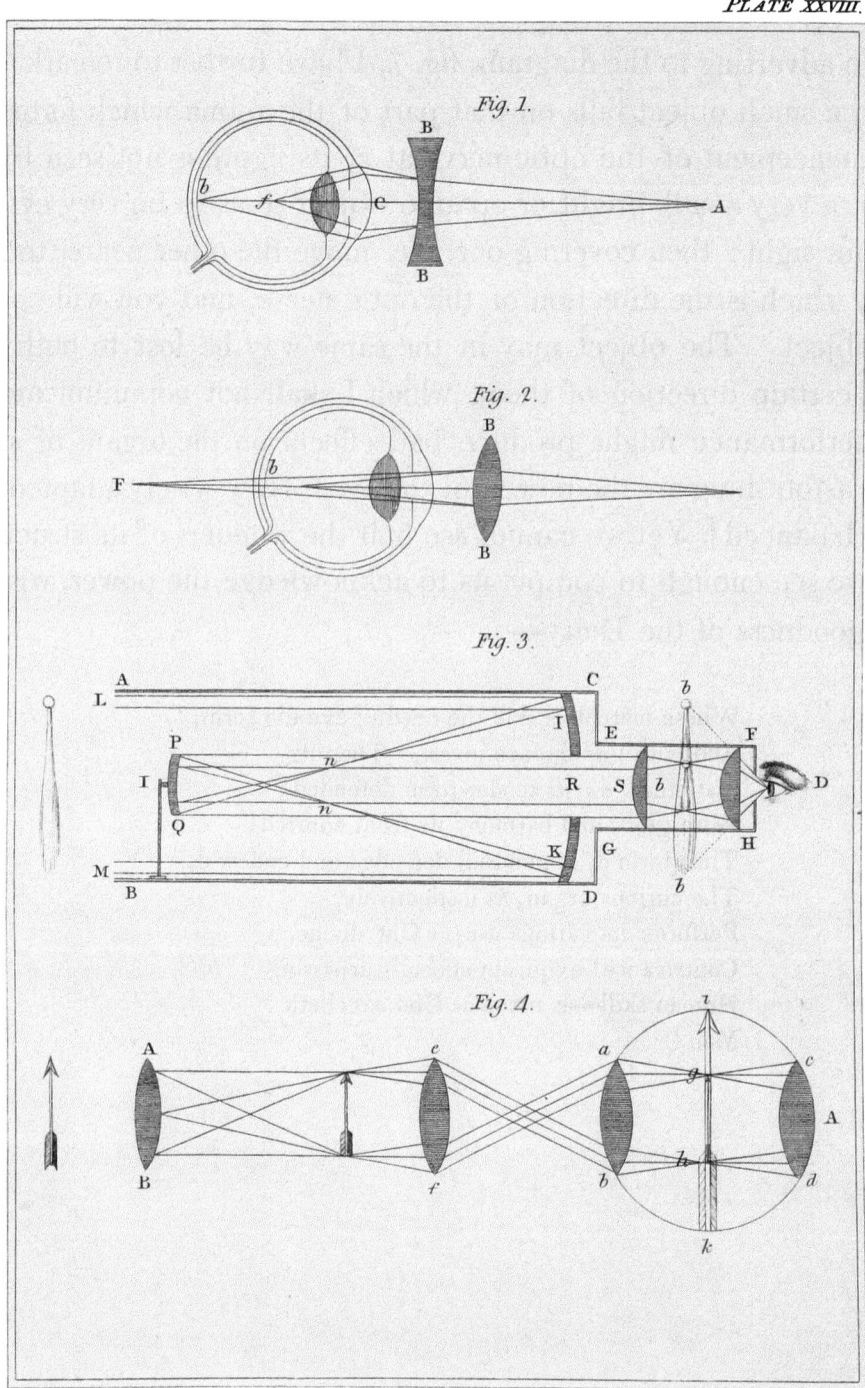

Fig. 1.

Fig. 2.

Fig. 3.

Fig. 4.

M.Bryan design.

H.Mutlow sculp.

PLATE XXIX.

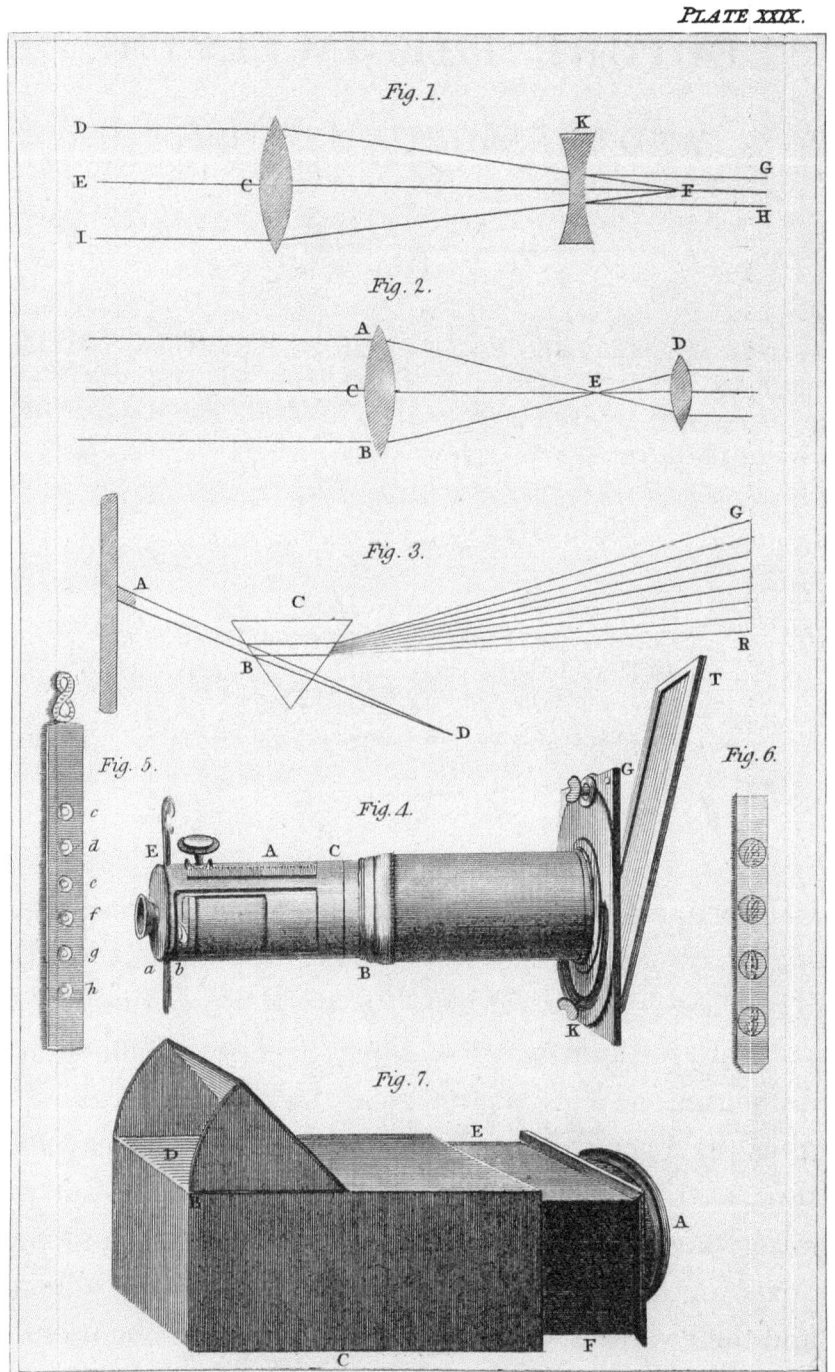

Fig. 1.

Fig. 2.

Fig. 3.

Fig. 5.

Fig. 4.

Fig. 6.

Fig. 7.

H. Mutlow sculp.ᵗ

LECTURE THE TWELFTH,

AND THE THIRD ON OPTICS.

VARIOUS OPTICAL INSTRUMENTS EXPLAINED AS TO THEIR CONSTRUCTION AND EFFECTS.—THE NATURE AND PROPERTIES OF THE PARTICLES OF LIGHT DISPLAYED BY THE PHENOMENA OF THE PRISM.—OF THE PERMANENT COLOUR OF NATURAL BODIES, &c.

Hail, Source of being! universal Soul
Of Heaven and Earth! essential Presence, hail!
To Thee I bend the knee; to Thee my thoughts
Continual climb; who, with a master hand,
Hast the great whole into perfection touched.   THOMSON.

EACH portion of the great Creator's wonderful works, when duly contemplated, equally excites our love and admiration; yet some of them, from the superior delights they afford us, naturally create a greater warmth of gratitude—a superior glow of affection: such delightful sensations I at this moment feel, while recalling to my recollection the beneficence of the great Creator displayed in my last lecture, and to be continued in the present one. But prior to investigating the wonderful properties of light not yet contemplated, I shall explain more fully some of its mechanical affections in lenses, and their various combinations. First, it will be useful to explain the manner in which the imperfections of sight, arising either

K K

from the protuberancy or flatness of the humours of the eye, are assisted by different glasses.

The near-sighted eye has the humours unusually convex; hence a concave lens is used to remedy the inconveniences arising from that circumstance; which beneficial office it performs in the manner I shall describe by means of the diagram, fig. 1, pl. 28. Let A represent a pencil of rays from an object falling on the cornea c, of a near-sighted eye. Were not the concave lens B B interposed, these rays would converge so much as to meet at f, without reaching the retina, and the vision of distant objects would be indistinct; for as objects are distinct only at the focus of parallel rays, the eye must either move nearer to the object, that these rays may meet in a focus on the retina b, or a concave lens must be used between the eye and the object to render the parallel rays divergent, to increase the divergence of the diverging rays, and to render the convergent ones less so, before they fall on the cornea c; this being effected by a glass of the required concavity, the rays then meet on the retina, as seen in the figure. When by age, or the habit of viewing distant objects, the humours of the eye become flatter, the images are thrown beyond the retina, as at F, fig. 2, pl. 28. This imperfection is remedied by a convex glass, B B, which rendering the rays more convergent in approaching the retina, causes them to meet on it, as at b. In this diagram, the focus of the rays at F represents the natural effect of the long-sighted eye; and the lines meeting at b, on the retina, the effect produced by the interposition of the convex lens B B. The latter defect may be remedied in young persons without the aid of a lens; but in extreme old age, merely withdrawing from objects will not be sufficient, because the eye may then not be capable of viewing objects in a weaker light.

Short-sighted persons view distant objects better as they advance in years; therefore, though in youth these suffer some trifling in-convenience, yet the future great benefit dependent on that youthful imperfection is a full compensation. But long-sightedness has not that consolation, for the defect increases with increasing years. But, thanks to investigation and the ingenuity of man, its most material disadvantages are surmountable.

Having treated sufficiently for our purpose on the subject of vision, I shall now explain the rationale of certain instruments which have been constructed on the discoverable formation of the eye, and the known principles that govern its operations and affections.

A Description of the Effects of Light in a reflecting Telescope.

The reflecting telescope, invented by Sir Isaac Newton, possesses a superiority over the refracting one, by being of a more convenient size, and producing a more steady and distinct effect. A two-feet reflecting telescope affords as great an advantage, and the same field of view, as a refracting one thirty feet long, and the light is more uniform in its effects by reflectors than by refractors.

Let A B C D, fig. 3, pl. 28, represent the large tube of a telescope, and E F G H the small tube containing the eye-glasses—so called from their situation near the eye. Suppose I, K, portions of a large concave speculum, to receive the images of objects. At the end of the tube A B a small concave speculum, I, is placed, which receives the image from I K, and transmits it to the eye.

In describing this telescope by a diagram, to prevent confusion it is usual to exhibit only two rays proceeding from the extremities of an object, as represented in fig. 3, by L M. Suppose L M to be rays proceeding from a comet, and falling on the extremity of the concave speculum ɪ ᴋ; they will cross each other at the foci ɴ, ɴ, of the angle they form with each other, and change sides; then crossing at the focus of parallel rays reflected from the speculum ɪ ᴋ, they will fall on the small speculum ɪ, in an inverted position: the rays from the object being received on the large speculum, and thrown thence on the small one, are returned towards the eye at ᴅ in the direction ᴘ ǫ; then passing through the aperture ᴙ in the large speculum, and the plano-convex lens at s, they converge so as to exhibit the figure of the object nearer to the eye, as at *b,* and upright. To view the object under a larger angle, a convex-glass is placed close to the eye, by which the image appears magnified or extended to *b b,* larger than it would appear to the eye viewed without the telescope, as is seen by the two images of the object.

We see objects magnified through telescopes, because the angle they make with our eye by the aid of this instrument, is larger than that they subtend at the distance we are placed from them. They are said also to be brought nearer to our eye, which merely signifies that they appear so much nearer to us by their images being brought to the focus of the lens.

Of refracting Telescopes.

The simplest of these instruments in its construction is that called the Galilean, from its being invented by Galileo. It consists of two lenses, one convex the other concave, as exhibited in fig. 1, pl. 29. Let c be supposed a convex lens, placed at the end of a telescope, and presented to an object, ᴅ ᴇ ꜰ, at such a distance that the rays fall parallel on the lens c; then passing through it, they would unite at the focus ꜰ were nothing intervening; but placing a concave lens to intercept them, before they meet in a point, at such a distance from the convex lens as will render the rays parallel, after passing through the concave one, these parallel rays exhibit the object at the focus of parallel rays, from the concave surface K, to an eye placed as at G H. Opera-glasses are constructed on this principle; the object is seen upright, because the rays are not suffered to come to a focus: but in the astronomical refracting telescope, the object is inverted. This instrument is constructed with two convex glasses. The object-glass, A B C, fig. 2, pl. 29, has its focus at E. The eye-glass, D, is much smaller, and more convex, than the object-glass; the latter being so placed in the instrument that its focus bisects that of the former at E. An eye placed at a distance from the eye-glass nearly equal to its focus, will see the object distinctly, and magnified in the proportion to the different lengths of the foci of the lenses.

You recollect that parallel rays falling on a double convex lens, and crossing each other at its focus, and then passing through another convex lens, go out of the second in a parallel direction, in the same manner as they entered the first lens; but that having crossed each other at the focus of the lenses, they represent the object inverted.

The inconvenience of seeing bodies reversed in refracting telescopes, may be remedied by the introduction of more eye-glasses, as thus: in fig. 4, pl. 28, let a b and c d be two additional lenses, added to the astronomical refracting telescope, of equal convexities with the eye-glass e f Then a b being placed at such a distance from c f as is equal to the diameter of the circle of which the convexity of e f is a portion, the rays, by the interposition of the lens a b, crossing each other at the focus of e f, will restore the figure to its true position, and bring it to g h; then being viewed by an eye placed at A, through the convex lens c d, it will appear upright. The degree of the magnifying power of a telescope is found by the proportion that the focal distance of the eye glass or glasses bears to that of the object-glass. But this calculation only expresses the increased diameter of the object; the area, or whole surface, of a body, is increased in proportion to its square.—Suppose the focal distance of the object-glass be eight times more than that of the eye-glass; then, for the whole increase of the area, we square the diameter, thus, $8 \times 8 = 64$: so that the object is magnified in that proportion. The images of objects viewed through refracting lenses often appear coloured, owing to some of the rays towards the edges of the lenses being unequally refracted, so as not all to meet at the focus.

Of the Camera Obscura.

To remedy the defect of seeing images inverted in a camera obscura, a mirror may be so placed as to exhibit the view of external objects in any required position. A box, fig. 7, pl. 29, having a mirror placed in an angle of 45°, as from B to C, will exhibit external objects in a perpendicular position to a person looking down on a ground glass placed horizontally, as at D; for the rays passing through a convex lens at A, fall on the mirror B C, and are reflected so as to appear at D.

If the camera be a room, the mirror, by different adjustments, will exhibit external objects in their true position on a white horizontal plane; or if the surface be made concave, to answer the convexity of the object-glass, the position of external objects will be more correct. When the object-glass is placed in a box, it has a sliding tube to regulate the focus of the glass to the distance of the objects viewed, as represented by E F, fig. 7. This instrument is sometimes fixed at the top of a building; in which case, the table that receives the objects in the room should be capable of adjustment by a screw, or some other means, to raise or depress it. A screw-table, with a concave surface of plaster of Paris, is the best, and also answers remarkably well for many astronomical purposes, such as seeing eclipses of the sun and moon, of which we can mark the progress much more conveniently than through a telescope in the usual way : besides, it does not fatigue the sight, and affords gratification to many persons at the same time. Celestial objects viewed in this way, will not appear so large as through a telescope. To magnify the object on the table, we use a small refracting telescope, screwed into a ball and socket, that permit our adjusting the instrument to the situations and distances of the objects we are observing; and also of seeing the spots on the surface of the sun, which are seldom discoverable by telescopes when used in any other way, on account of the necessary obscurity of the glass used to transmit the rays of the sun to the eye. The transits of Mercury and Venus, and all other celestial phenomena, are pleasingly exhibited by the same apparatus.

The Magic Lantern.

This instrument is too well known to require a representation; yet I shall describe it, as by its application to astronomical and other natural subjects, it may be rendered a very desirable apparatus for the instruction, as well as the entertainment, of young people: for by sliders calculated to exhibit certain interesting subjects, natural phenomena may be agreeably illustrated; such as the apparent retrograde motion of the planets, the phases of the moon, the causes of the vicissitudes of seasons and eclipses of the sun and moon, the transit of a planet over the sun, and the stars forming different constellations of the heavens: this application of the magic lantern renders what is usually considered merely a trifling amusement for children, an important and interesting object of attention and information to young students and philosophers.

This machine consists of only two double convex lenses; one of which, being extremely convex, renders the light from a lamp placed behind it very dense at its focus, and thus shows the object distinctly. This light passes through the glass on which the object is painted, and is received bright on the other convex lens, which has a long focus, and is placed in a sliding tube, to adjust the focal distance. The objects are thrown on a white screen, where they appear very bright, the rest of the apartment being dark; and the illumination will be increased, if a concave mirror be placed behind the lamp. The images are made to appear in their natural position, by inverting the sliders on which they are painted.

Of the Solar Microscope.

Microscopes are made of various forms and capacities, but I shall only describe the one which comprises all the advantages of the rest, and also possesses one quality exclusively, which is that of exhibiting objects in a darkened chamber, so that they may be viewed without fatiguing the eyes; from which use it is called the solar microscope, and by certain additional appendages may be rendered a lucernal one. I shall explain the general properties of the *lenses* used in microscopes, so far as will serve for comprehending the rationale of all the others expressed by different names; without going into a tedious detail of the applications of their various qualities and adjustments, under all the forms of this instrument, which are rather mechanical, than physical considerations.

Objects that at about six inches from our eye are too minute for inspection, may be viewed distinctly by the assistance of glasses: hence the instrument constructed to exhibit such very small objects is called a microscope. To bring the object nearer to the eye, a convex lens is used ; for as the object appears at the focus of this lens, of course it forms a greater angle at the eye.

The microscope consists of a tube of brass, A, fig. 4, pl. 29, which screws into a socket, B ; a convex-glass is placed at the end, c, of this tube, to bring the object nearer to the eye when it is used as a simple microscope. The axis of this lens is exposed only when very large magnifiers are used, for which purpose a concave plate of brass is screwed over the lens, to prevent the dispersion of the rays of light that are to illuminate the object. At the other extremity of the tube, at E, a space is left, between two plates of brass, a, b, to receive the objects to be viewed. A hollow tube is sometimes used when living objects are examined, but which I never employ in the solar application of this instrument, as it must torture a poor insect to subject it to the intense heat of the sun. In fig. 5, c, d, e, f, g, h, are six different magnifying powers, fixed in a brass slider. Fig. 6, is a slider of ivory, supplied with pieces of talc, to inclose the objects to be viewed.

To examine an object, we place the ivory slider between the flat brass plates, as represented at E, and the ivory one containing the object close to it. The side of this slider that has the brass rings, should be placed farthest from the eye, to suit the focus of the glass.

To render a simple microscope a solar one, we require an additional apparatus, like G K, into which we fasten the tube already described. Screwing the plate G K, which supports a mirror, I, to a window shutter, in the position represented by fig. 4, we are enabled to throw the sun's light on the object in the microscope; then placing a white screen opposite to the instrument, an enlarged image of the object is painted on it.

The multiplicity of optical instruments precludes even a general description of them in this lecture; nor is it necessary, for the foregoing explanations of some of them, added to the known properties of lenses and mirrors, and the effects of light, enable us, on inspection, to understand the construction and operations of all others presented to our observation.

We are now far advanced in the consideration of light and the science of optics; having, by ocular demonstration of certain results, inferred with certainty many important facts: such as, that the particles of light are inconceivably small, and move in a rectilinear direction with astonishing velocity—that a ray of light, radiating from a centre, diverges in its progress—that the density of light at certain distances depends on its density at the radiating point and its distance from it, and this difference being also in proportion to the squares of its distance from the luminous point—that the angle made by a ray of light in its reflection, is always equal to its angle of incidence; and hence, when the angle of incidence is found, the angle of reflection is likewise ascertained—that concave mirrors collect parallel rays, and cause them to meet in a focus by reflection; and that the focus of a concave mirror is at the same distance from its surface as the focus of a convex lens—that the heat and light of a luminous body reflected from a concave surface, are as much increased at that focal point, as that point exceeds the surface of the lens; the same as happens in regard to the surface and focus of a convex lens by refraction, which causes the rays of light at the focus of very large concave mirrors and convex lenses, by being greatly accumulated at their foci, to burn almost all bodies subjected to their influence. We have also contemplated the curious organization of the eye, so far as its optical effects are known; and discovered, that the construction of optical instruments depends on the known properties and capacities of the coats and humours of this useful and ornamental organ of the animal creation.

On Colour.

But the sublimest evidences and most beautiful effects of the particles of light yet remain to be considered; namely, the different sizes of these particles, with their various impressions on the organs of sight, and their individual characteristics of colour.—

———————— Who can paint
Like Nature? Can imagination boast,
Amid its gay creation, hues like these?
Or can it mix them with that matchless skill,
And lose them in each other, as appears
In every bud that blows?   THOMSON.

For all these effects, under Providence, we are indebted to the bright luminary of day; which thus adorns and paints the face of nature with different graces, according to the capacity of substances to imbibe or reflect its beautiful emanations. The sublime Newton has furnished us with the clearest evidences of these effects; having shown, by unequivocal experiments, that the rays of light consist of particles differing in colour, though, by a due mixture and perfect combination, they exhibit a pure white light.

What an astonishing exercise of human reason is here displayed! What delight must he have felt when, by his soaring and penetrating genius, he made this astonishing discovery! conceived by his own amazing powers of perception, energy and judgment! I can conceive no task more difficult for the genius of an artist to execute, than to paint the countenance of this great man when he identified these facts by his experiment with the prism. For who can ex-

press the combined passions that at one instant possessed his mind
—joy at the success of his experiment—astonishment at the brilliant beauty, and admiration of the divine harmony displayed in
this interesting spectacle?

The phenomenon of the prism is supposed to depend on the
different sizes of the particles of light, as demonstrated by the laws
of motion and the resolution of forces. The colours exhibited by
the spectrum, prove that the particles of light consist of seven
different colours.

> First, the flaming red
> Springs vivid forth ; the tawny orange next;
> And next delicious yellow ; by whose side
> Fall the kind beams of all-refreshing green.
> Then the pure blue, that swells autumnal skies,
> Ethereal blaze ; and then, of sadder hue,
> Emerge the deepen'd indigo, as when
> The heavy skirted evening droops with frost ;
> Whilst the last gleamings of refracted light
> Died in the fainting violet away.   THOMSON.

Who can doubt of these being nature's resplendent robes ; imparting
grace and genial warmth to the various forms of nature?

Let us now contemplate the instrument contrived by the intelligent Newton to authenticate his divine theory of light and
colour. Fig. 9, pl. 26, exhibits a triangular prism, which, by the different degrees of its refractive power, causes the rays of light, according to their various sizes or colours, to pass through different portions
of its surface.

Let A B, fig. 3, pl. 29, represent rays of the sun's light passing into a dark chamber, through a small perforation in a window-shutter, and falling on a triangular prism; then will these rays be divided from each other, and arranged according to their respective classes, by the refractive power and angular form of the prism. The red rays being the least refrangible, the intervention of the prism will turn them the least from their straight course, while the violet are the most bent and turned aside by the same cause. These rays would have proceeded in the direction from B to D, had not the prism intervened.

The refracted rays will exhibit, on a white screen placed to receive them, an oblong coloured spectrum. In the diagram, the parts between the lines are the rays: R represents the red, which are supposed to be the largest particles of light, and therefore the least affected or turned aside. The violet, being the most refrangible, appear at G; and all the intermediate rays arrange themselves according to their classes and supposed sizes, and thus exhibit their respective colours and gradations.

Many curious experiments were made by Sir Isaac Newton to establish these facts, and to prove that the colours of the different rays were unalterable by reflection or refraction, and that a due mixture of all the colouring particles was necessary to produce a perfectly white light. The small aperture for these experiments should be about a quarter of an inch in diameter, and the room perfectly dark. If we make a hole with a pin in a piece of paper, and let only one of the coloured rays pass through it, the colour of that ray alone will be seen on a white screen placed behind the perforation: if we then let the reflected ray from the coloured spot pass through another prism, it will still retain its colour, and also its form, appearing perfectly round. Any object held in the separated ray will appear tinged with its colour. Lastly, if you look at this coloured spectrum through another prism, it will not suffer any change in its colour, form, or size. From these experiments it is with reason inferred that the rays of light are of different colours, and that in their homogeneous or separated states these colours are unalterable.

The next experiment will show us, that the white appearance of light results from a due mixture of all these colouring particles. Place a convex lens to receive all the coloured rays of light when separated by the prism; these passing through it in their separated state, and afterwards uniting at its focus, will exhibit pure white, like the light of day. If between the prism and screen we place a card with a small perforation, suffering all except one of the coloured rays to pass to the convex lens, the spot at its focus will not be pure white, because deficient by one portion of its due mixture to produce that appearance. Again, if we suffer any two of the primitive colours to pass through the perforation in the card, the spot at the focus of the lens will exhibit a colour resulting from a mixture of these two colours. All the phenomena of different lenses and prisms may be seen by a stream of light in a dark chamber. A very dark apartment is necessary for these experiments, for a small portion of extraneous light will diminish the effects.

The beautiful appearances of the different colouring particles of light are exhibited by a jet of water, fig. 8, pl. 10. Placing yourself between the sun and the fountain, the globular particles of water thrown up by the jet, fig. 2, pl. 10, falling in a spherical form, will, to a person placed at a due distance, and in a proper position, which may be found by trial, have a beautiful appearance like the rainbow. This phenomenon arises from the different refrangibility and positions of the portions of a sphere which form it, and therefore may be exhibited by a globe of water, which, possessing different powers of refrangibility, exhibits the colours of the rainbow to the eye of an observer properly placed for the observation.

Fill a glass globe with water, and, suspending it by a string, suffer the rays of the sun (when that luminary is near the horizon) to fall upon it; then placing yourself in a proper position in respect to the sun and the globe, namely, directly in the line between them, you will perceive all the colours of the rainbow.

Having established the most interesting facts relating to the nature and properties of light, we will next consider some other inferences which have been deduced from them. The ancient philosophers imagined colour to be a quality of bodies independent of light; but later philosophers have proved that it results from different modifications of light, and the constitution and construction of bodies, which dispose them to reflect some colours, and to imbibe and absorb others. This opinion is founded on the solid and consistent reasoning of Sir Isaac Newton; who also conceived that the colours reflected from transparent colourless bodies resulted from their form, which, by refraction, divided the colouring particles of light from each other, and thus manifested their individual properties. As the images of objects are produced on the retina by actual impressions, he supposed that the effects arising therefrom were proportional to their sizes; the smaller particles exciting the shortest vibrations in the retina on account of the velocity of their motion and their minuteness, and the largest particles exciting the longest vibrations in the retina, for the contrary reason. He also conceived that the rays reflected from

coloured bodies produced similar sensations on the retina; and that
these colouring particles, both by reflection and transmission, excited longer or shorter vibrations than others in proportion to their
difference of size and the quickness or slowness of their percussion;
in the same manner, we may suppose, as the impressions of sound
are conveyed to the ear, and produce different sensations on that
organ. This great man also supposed colour reflected from permanently coloured substances, to arise from their individual disposition to reflect this or that sort of rays more copiously than the
rest; and that these rays propagated their impressions by exciting
sensations on the retina, and thence to the brain. If it be not thus,
how is it that some colours, in looking on them, fatigue the eyes
more than others?

The white appearance of the light of the sun is produced by a
copious reflection of all the coloured rays, and the velocity of
their motion. Objects on our earth appear white when they reflect all the rays of the sun; but blackness is produced by an absorption of the incident light, which, being stopped and suppressed
in the black body, is not reflected outwards, but inwards, and refracted within the body till it is stifled and lost.

The sublime investigations of Newton resulted from the observation of the *calorific* appearance of the rays of light in soap bubbles, which led him to observe these effects in substances of different
forms and densities. In this investigation he determined that the
diamond, among colourless bodies, from the density of its nature,
both refracted and reflected the light more copiously than any other
transparent substance. As light will pass through the least particles
of bodies, their opaqueness in the gross must arise from innumerable

reflections and refractions in their interior parts, arising from the number and situation of the spaces filled with different mediums, and in the refracting particles which compose them.

Fill the pores of opaque bodies with any substance of nearly the same density with themselves, as the interstices of paper with oil; and the whole will become transparent. As filling the pores of an opaque body makes it transparent, so emptying the pores of transparent bodies, or rather separating their parts, renders them opaque. The interstices of bodies must not exceed a definite size, to render the bodies themselves opaque and coloured. Hence it is that water, glass, &c, are not opaque, but transparent; their particles and pores being too small to cause reflection at their common surfaces.

The transparent parts of what we call opaque bodies, according to their different sizes and situations, may reflect rays of one colour and transmit those of others; and this forms the ground of all their various colours. The effects arising from the different reflections produced by the transparent parts of bodies, are evident in the variety observed in different views of the plumage of some birds. That these effects are produced by the reflection and refraction of light from their surfaces is rendered evident by wetting the feathers, when their reflections will be less vivid, and the colour more uniform.

The skin of the cameleon is transparent, and of a yellowish-red ground, covered with small smooth protuberances of a bluish colour. This skin is endowed with the faculty of expanding and contracting itself, which causes these colours to vary in their appearances; seeming sometimes red, and at others blue; and when the yellow

M M

rays of the ground mix with the blue of the protuberances, the skin appears greenish.

The red and orange rays of light, by the momentum of their force, are transmitted to a greater distance than the fainter rays; for some of the latter in proceeding towards our earth from the sun are stopped in their passage by our atmosphere, and are reflected from other bodies. This circumstance explains the blue shadow of bodies; and the blue colour of the sky is likewise supposed to be produced by copious reflection of the blue rays by the atmosphere. The sun's horizontal light is often deeply tinged with red and orange; this usually announces fine weather, as it denotes a clear state of the atmosphere through which the light is transmitted, which permits the passage of the stronger rays when the fainter are withdrawn from our view. It is thought that these rays, by being thrown into the shadow of the earth, cause the moon when eclipsed by that shadow to appear of a reddish colour. This conjecture is very probable, as the degree of this colour varies in eclipses according to the extent of the earth's shadow at the time. Sir Isaac Newton's theory of the permanent colour of opaque bodies is this: when the colouring medium of bodies is so dilute that the light may be transmitted through them, their colour appears vivid; but when the tinging matter is more densely diffused in them, they appear black. From which we may infer, that the colouring medium of vegetables does not reflect light, but only their solid particles, which have the property of reflecting some of the rays of light, and of imbibing others.

Sir Isaac Newton conceived that the colours of vegetables might be produced by the light reflected from their substances, and transmitted through the colouring medium which resides on their surfaces. To identify this idea, he extracted the colouring particles from plants, by digesting them in rectified spirits of wine; and found, after the colours had been totally imbibed by the spirit, that the leaves and flowers were unaltered as to texture, but appeared in some cases perfectly white, and in others nearly so. The spirit exhibited either no colour or very little indeed: but when floated on white paper, it exhibited the colour it had imbibed; displaying the colour only by transmission, not by reflection. This confirmed his opinion, that the colouring medium does not of itself reflect light, but, being transparent, is reflected from the solid parts of bodies. He next considered the colouring property of animal bodies, and found the effects were similar, in regard to the manner in which the colour was produced and reflected, to those in vegetables. White paper shows colour when the colour is laid on so thinly that the white can act through it, but not else. The coloured substances of the mineral kingdom belong principally to two classes, earths and metals. The former, when pure, are all white; and their colour may arise from phlogistic or metallic mixtures. Flints owe their dark colour to the state of fire in them; when this fire is struck out, the stone appears white, by the loss of its inflammable principle. We may suppose that coloured gems derive their appearance from metals, as they are imitated by tinging glass with inflammable metallic matter. The particles even of transparent bodies are quite white, as may be seen in pulverised glass: therefore coloured gems, like other substances, exhibit their colours by the transmission of the colouring medium from a white ground.

The earthy particles which form the solid parts of different bodies, and are the ground-work of colour, vary in size and density: accordingly they reflect the rays of light with a proportionate force; or rather with a force proportioned to their difference of density in

respect to the medium of the air.   Transparent colourless bodies,
such as air, &c. reflect all the rays of light; hence they appear
white.   The mediums of bodies which cause them to reflect all, or
some, of the rays of light, depend on their surfaces, as well as on
their constitutional medium.   Of this we have evidences in metals
called white : for these substances do not appear white unless their
surfaces are rough, as in that case only there are interstices on their
surfaces sufficient to contain the medium which reflects light; and
when these are accumulated, a white and vivid light appears. Such
bodies as have their interstices so disposed that the light can pass
through them, are perfectly luminous; but those not so constructed
do not afford a free passage to the light, therefore we see only those
rays which are reflected from their surfaces.

It is probable that the constitutional medium which causes those
natural inflammable substances, that are of a dark colour, to imbibe
the rays of light, imbibes fewer or more of these rays according to
the state of fire in such bodies.   Sir Isaac Newton found that inflammable substances possessed a much greater degree of refractive
and reflective power than those which were not inflammable: from
which he drew the conclusion, that the diamond contains a larger
proportion of the inflammable quality; and also because when excited it exhibits in the dark the character of light or fire.

Having contemplated the solar rays under different modifications of heat, light and colour, we will next consider the effects of
that light emitted from some substances in the dark, called phosphoric.

Phosphori are divided into three classes, to distinguish those
bodies that require previous exposure to the sun or fire, such as

earthy substances, to become luminous; and such as do not, as rotten substances, and the phosphoric skins of fishes; and lastly, such as require excitation to cause an emission of their light, namely, sugar, salt, water, &c. Those bodies which become phosphoric by excitation, may have this quality produced by different powers impressed on them, such as attrition, agitation, heat, the free admission of air, and being exposed to the external light. Most bodies that can bear that friction which is necessary to produce their latent fire, will become phosphoric by attrition.

Agitation agrees with liquid substances, as sea water; gems will become phosphoric by the application of heat; the free admission of air produces a flow of light in phosphorics prepared from earths, &c. It has been ascertained, that there are but two substances that do not emit light when brought into total darkness, after having been exposed to a strong light; namely, water and metals; thus we find all other bodies have not only the power of imbibing light, but also, when placed in proper circumstances, of emitting it.

It is said that a hand, after exposure to the bright light of the sun, being introduced into a closet made so dark that no reflection or glimmering of light appears, by means of a contrivance that admits of the ingress of the hand without introducing any light but what the hand conveys, will appear luminous in the dark; that white paper possesses this property in a very high degree, for if an opaque substance be laid on it before the paper is presented to the light, on removing the opaque body into total darkness, a part of the paper will appear dark, exactly of the figure and size of the interposed body.

Though water in its fluid state is incapable of emitting light, yet when condensed into ice or snow it assumes it like other bodies.

We now understand the cause of the light afforded by snow during the night;—its being exposed for many hours to the bright light of day, and consequently imbibing a great quantity of the sun's rays, and afterwards emitting them in the absence of that luminary.

Phosphoric bodies are all those which, on being excited, give a faint light that is visible only in the dark; or those that, after being exposed to a bright light, emit the light they have imbibed. Phosphoric and phlogistic bodies agree in containing a quantity of light which is not in a perceptible state of heat. Those bodies are in some respects analogous, as they emit light on the same principles, so far as depends on the luminous matter being set at liberty by the operation which renders them luminous; but the manner in which it is liberated differs in each from the other, as does also the cause which retains the luminous matter in these different substances.

We perceive no alteration in phosphoric bodies after they are deprived of their shining quality; but phlogistic bodies appear very differently after they have lost their luminous property: for phlogiston is the inflammable part of bodies, but phosphorus is only an emission of the light previously imbibed.

From our observation and examination of fire or light, we are induced to acknowledge its characteristic of materiality. This solar substance appears under the different modifications of fire, light and electricity, according to the situation in which it is placed; yet the same properties are discoverable in each of these, which determines their identity

Throughout our investigation of light or fire, we have perceived that it is a power which causes motion in bodies, and that without-its animating influence all nature would become languid, sicken and decay: that it is equally as essential to plants and vegetables as to man; and determines the state of different substances: that it is a principle more essential even than air to the existence of plants and vegetables; and to the solar rays they are also indebted for their beautiful and gay attire. Flowers and vegetables, excepting those that are phlogistic, lose their variety of tints when deprived of the solar influences, as we perceive to be the case with endive, celery, &c.

Some plants show a greater sensibility to the solar light than others, by expanding their leaves and turning to the sun. That it is merely the light, not the heat, of the sun which causes these effects, we may be convinced by experience and observation; for plants placed in a room in which there is a fire, will turn constantly to the lighter though that should be the colder side. We may suppose also that fruits and flowers derive their flavour and perfume from the solar rays; as we find these most fragrant and highest flavoured, when produced in climates where the light is the most intense.

Animals living under ground are of a dull colour; and persons look sickly and pallid who are constantly immured in dark apartments. The same effect is seen in wood, which has little or no colour before it is exposed to the light, but when receiving the solar rays its substance becomes highly coloured.

From the whole of Sir Isaac Newton's theory, after having read and digested it with my closest attention, I infer, that he considers the permanent colours of bodies to be dependent on some essential quality, varying in each species, which causes different bodies to absorb some coloured rays, and emit others; and that as these properties are variously modified in different substances, one absorbs all but the red-making rays, another all but the blue, and so on, through the whole scale of colours; and thus they reflect different colours to our sight. If these colouring particles be of different sizes, which we have good cause to believe, they must produce different sensations on the retina: but the property of reflecting colour exists in the particles themselves; so that the appearance of colour arises accordingly from a real, not an imaginary, quality of bodies. And speaking of the sensations produced on the retina by different colours, Sir Isaac Newton meant to convey the idea of this being produced by the essential properties of bodies, and of the different coloured rays of light. He supposed also that the coloured appearances produced by colourless transparent bodies arose from the separation of the rays of light by refraction.

How charming are the evidences of the Deity we have just been contemplating! How unequivocal the effective energies of light! The variety, multiplicity and beauty displayed in this subject, produce such a quick succession of pleasurable sensations, that it is impossible to give either individually the preference. But the great Cause of the effects perceived, rising supremely conspicuous above them all, claims and receives our first attention—our most exalted and concentrated admiration, love and gratitude!

> By swift degrees the love of nature works,
> And warms the bosom, 'till at last, sublimed
> To rapture and enthusiastic heat,
> We feel the present Deity.   THOMSON.

PLATE XXX.

Fig. 1.

View of the Proportional Magnitudes of the *PLANETARY ORBITS.*

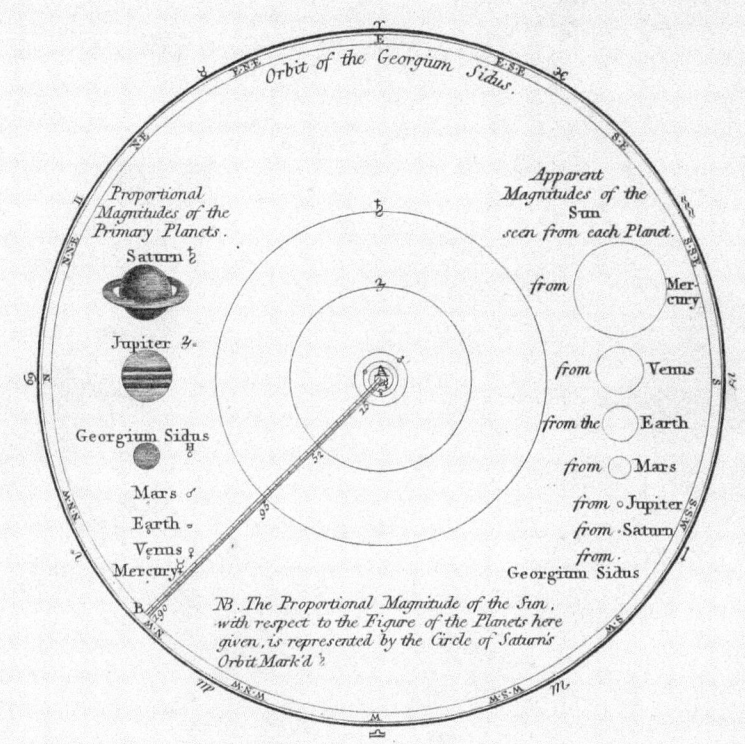

Fig. 2.

Right Sphere.

Fig. 3.

Parallel Sphere.

Fig. 4.

Oblique Sphere.

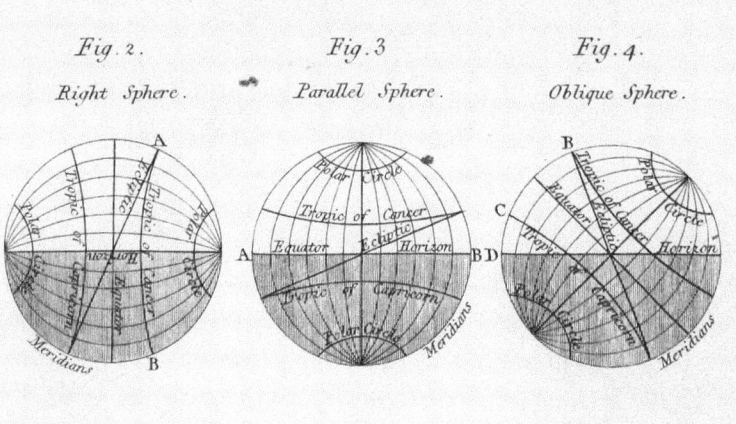

H. Mutlow sculp.

PLATE XXXI.

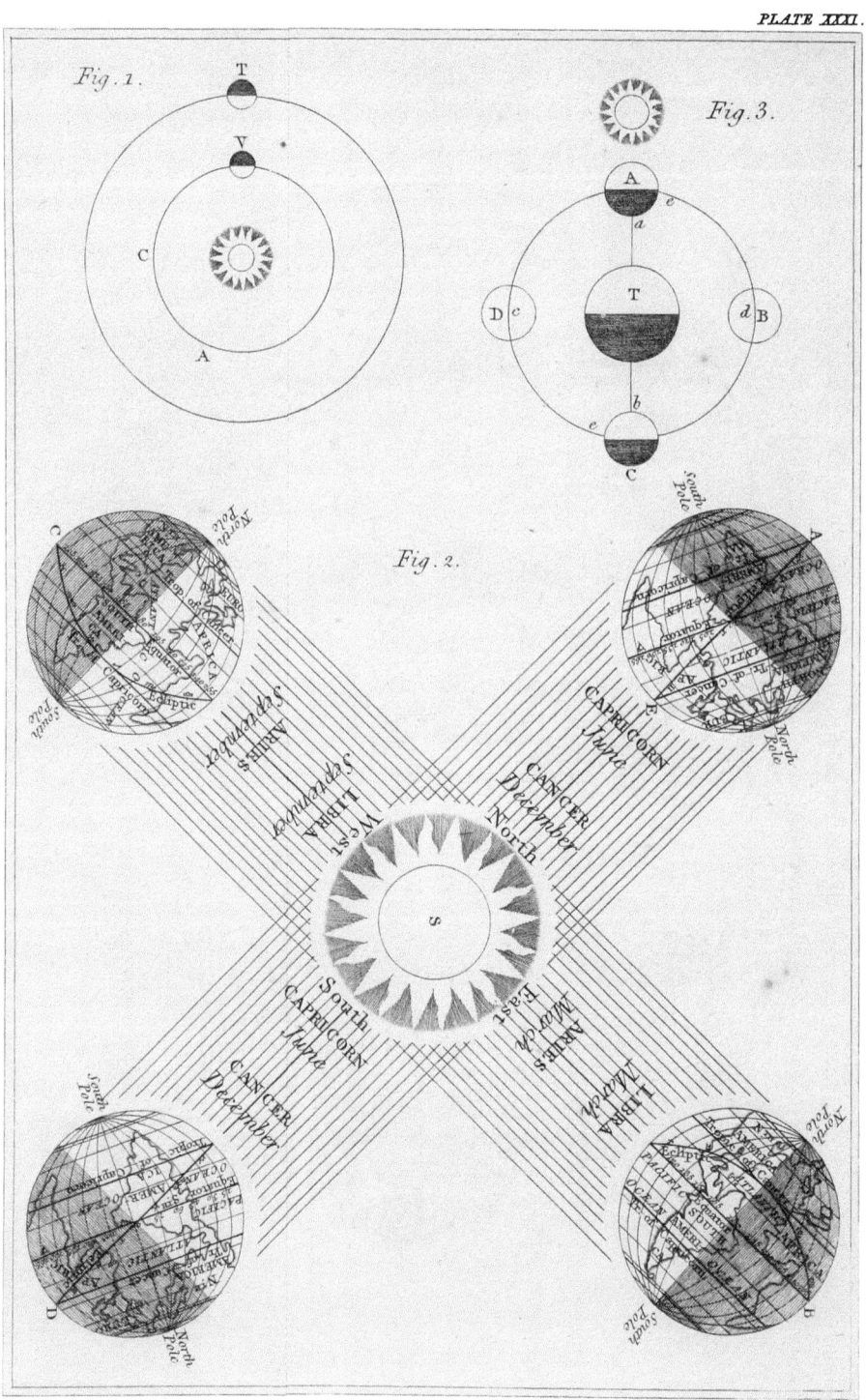

Fig. 1.

Fig. 3.

Fig. 2.

M.Bryan design.

H.Mutlow sculp.

PLATE XXXII.

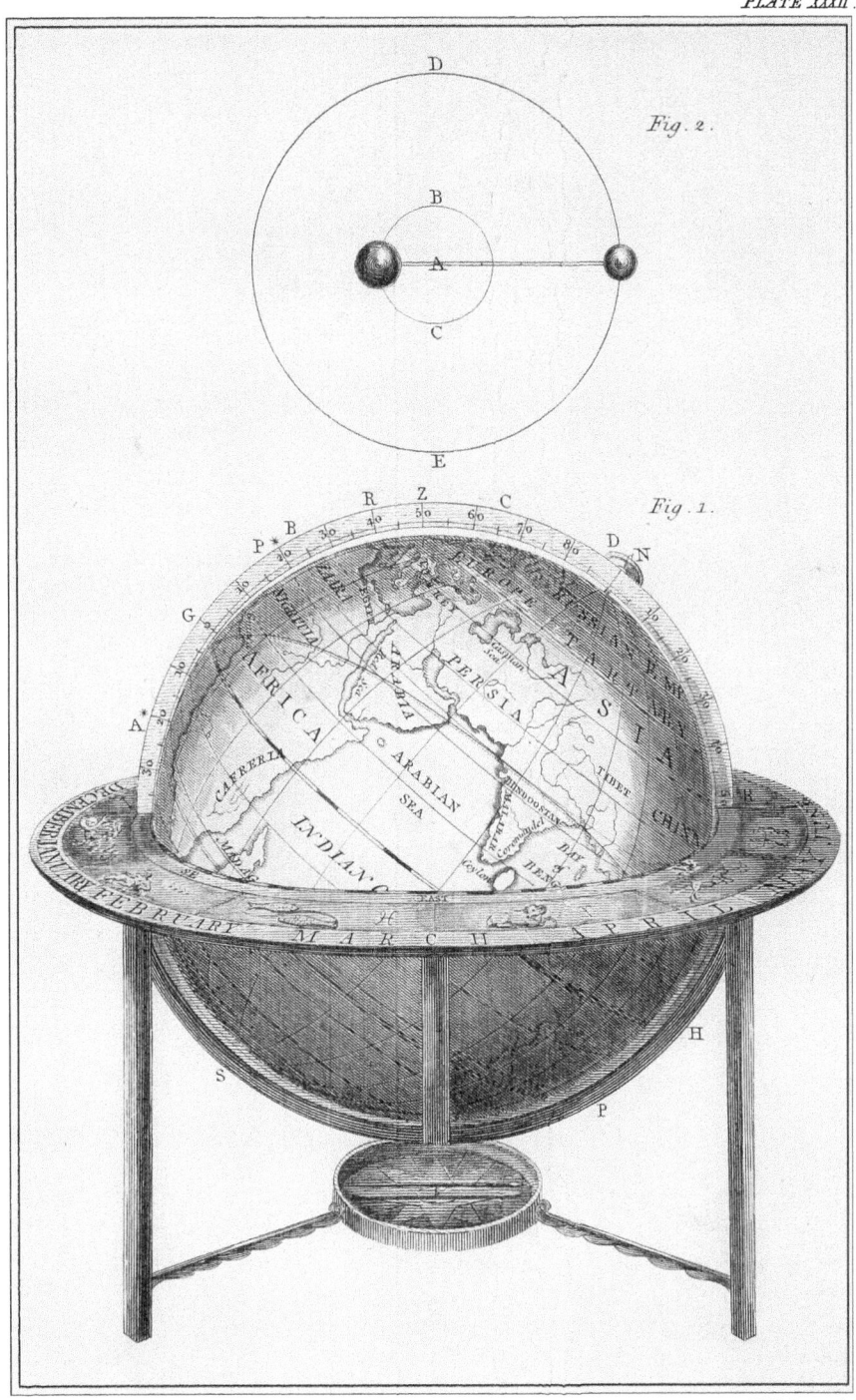

Fig. 2.

Fig. 1.

T. Noble del. H. Mutlow sculp.

PLATE XXXIII.

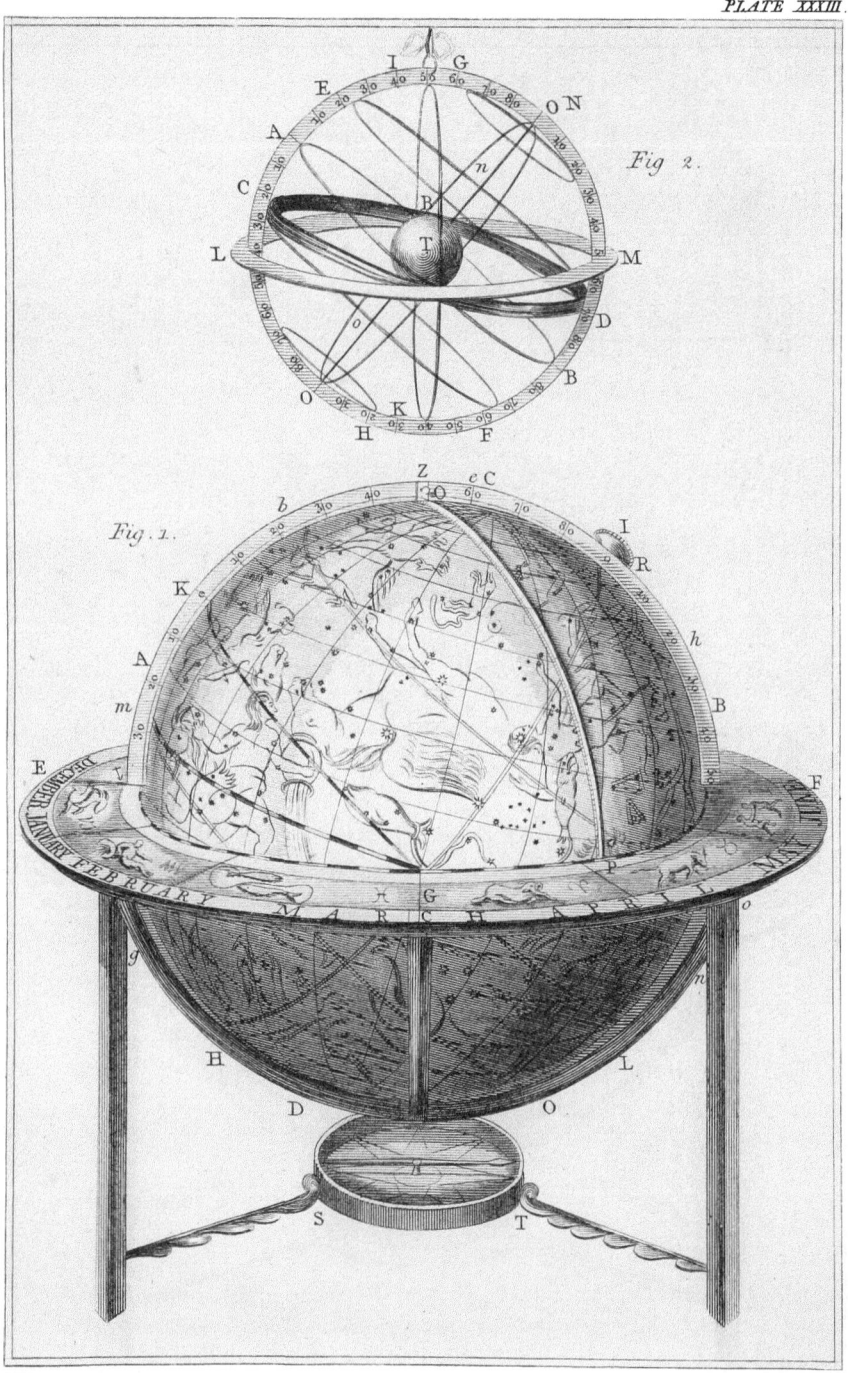

Fig 2.

Fig. 1.

T. Noble del.

H. Mutlow sculp.

PLATE XXXIV.

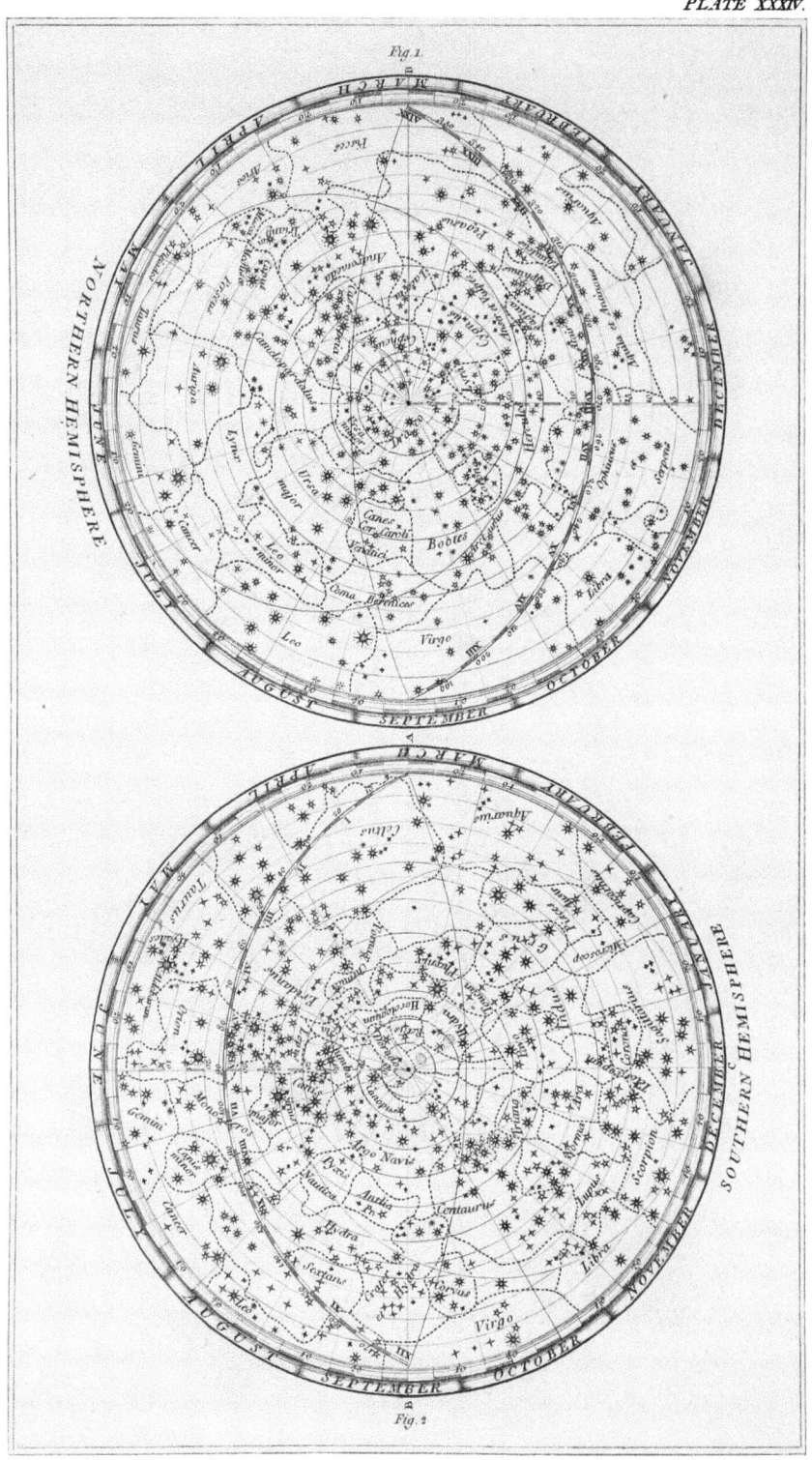

Fig. 1.

NORTHERN HEMISPHERE

Fig. 2.

SOUTHERN HEMISPHERE

T. Noble del.

H. Mutlow sculp.

PLATE XXXV.

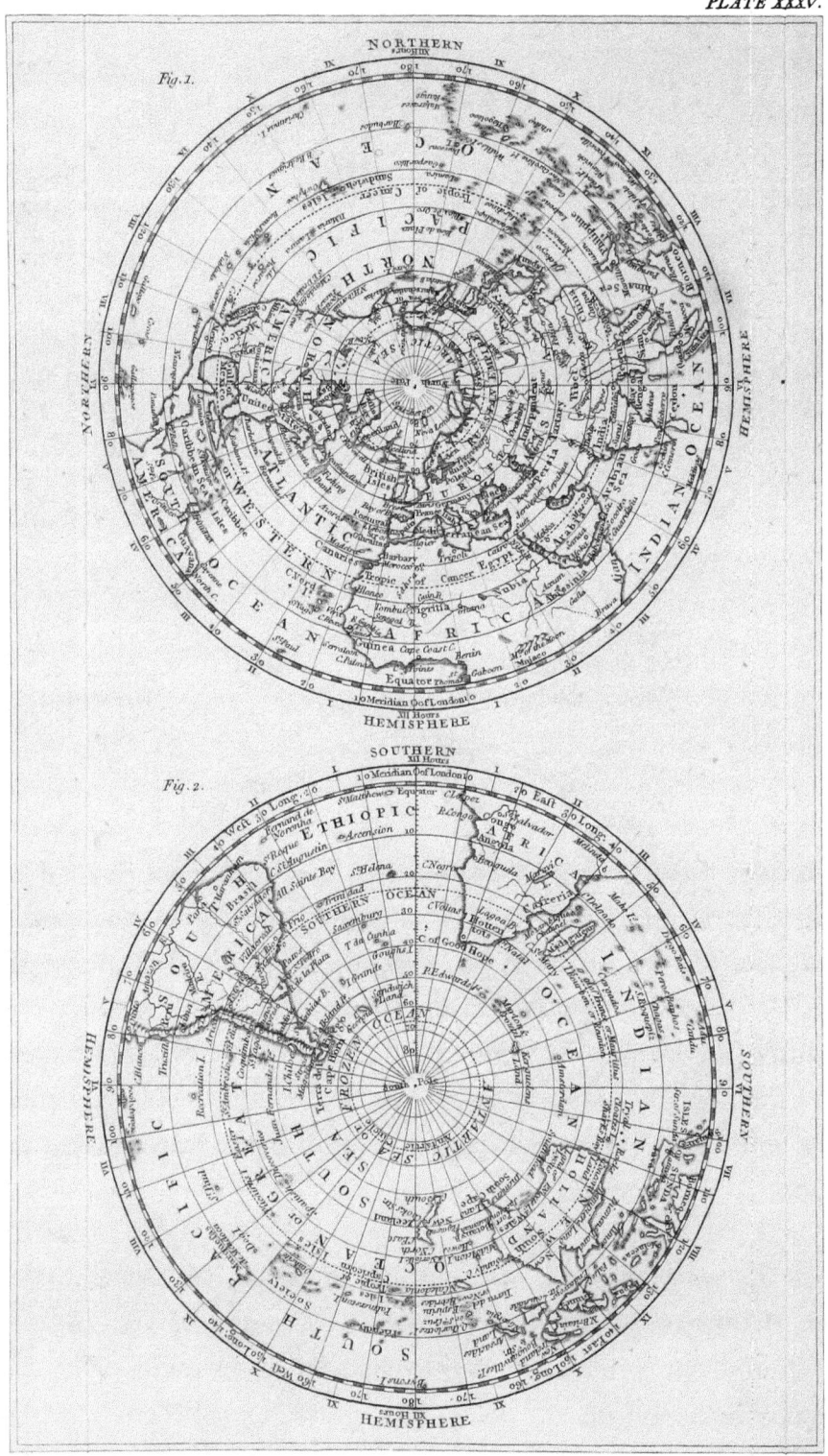

Fig. 1.

NORTHERN HEMISPHERE

HEMISPHERE

Fig. 2.

SOUTHERN

HEMISPHERE

Maslen sc. Russel Co.F.

LECTURE THE THIRTEENTH.

ON ASTRONOMY.

———————

THE SOLAR SYSTEM EXPLAINED; ALSO THE CIRCUMSTANCES WHICH HAVE
PROVED THE ROTUNDITY OF THE EARTH, AND THE SUN BEING THE CEN-
TRAL BODY OF THE SYSTEM; WITH THE OTHER PHENOMENA RELATING TO
THE REVOLUTION OF THE PLANETS, AND THE CAUSES OF ECLIPSES, &c. &c.

———————

Inspiring God! who, boundless spirit all
And unremitting energy, pervades,
Adjusts, sustains and agitates the whole.   THOMSON.

———————

SO grand, beautiful and sublime is the whole scheme of the
universe, that it requires the association of all the most elevating
ideas, to raise the mind to a pitch of thought capable of conveying even the weakest impression of its astonishing excellence! yet
the assimilating power of science enables us to calculate many
of its sublime effects, and to view and understand its resplendent
beauties, and most powerful energies, with ease, satisfaction and
conviction.

Aided by mathematics, we venture to speak with certainty of
the sizes, distances, periods and motions of some of the heavenly
bodies, though far removed from our familiar inspection. The solar
system, exhibited by fig. 1, pl. 30, was first established by Pytha-

goras; and since revived by Copernicus, after the exuberance of genius had been corrected by the infusions, and modified by the restrictions, of science. That the motions of the heavenly bodies excited the attention of the earliest ages we may readily believe; for the necessities of human nature must have naturally led men to contemplate the aspects of the sun and moon for different times and seasons, in order to regulate the affairs of agriculture and domestic employment. The results of these early investigations excited an increasing curiosity in the breasts of intelligent men, and led them to contemplate the fixed stars, the influence of which was then much considered and accredited. But science being incompetent to enable men to ascertain either the sizes or distances of the stars, they are only distinguished by their different apparent magnitudes; and by being grouped into constellations, and characterized by names, either of particular observers of these beautiful luminaries, or adapted to different events in profane history. This arrangement was particularly useful to navigation before the use of the magnet was discovered *.

In the centre of the solar system, fig. 1, pl. 30, is placed the sun, like the father of a family, surrounded by bodies dependent on his emanations, called planets; one of which is our earth. Mercury, ☿, is situated nearest the grand luminary; next Venus, ♀; then our Earth, ⊕; Mars, ♂; Jupiter, ♃; Saturn, ♄; Herschel or Georgium Sidus, ♅.

Three other planets, lately discovered, have not yet been introduced into astronomical tables; yet I must not neglect mention-

* For an historical account of the progress of astronomical knowledge, see my Astronomy, octavo edition.

ing them. Of these, the two first-discovered are called Piazzi and
Olbers, after the names of their discoverers; or, as they are otherwise called, Ceres and Pallas. The former of these planets was
discovered on the first day of the present century—namely, January
1, 1801—by M. Piazzi, astronomer, at Palermo in Sicily; and the
latter on the 28th March, 1802, by Dr. Olbers, astronomer, at
Bremen in Germany. Both of these planets appear extremely
small, like telescopic stars of the seventh or eighth magnitude.
They move in orbits between those of Mars and Jupiter, and in
some parts of their tracks they approach very near to each other;
and, what is singular in our observations of these planetary bodies,
their orbits cross each other;—the planet Olbers coming nearer to
the sun than Piazzi in the perihelion, or near part of their orbits;
but going off to a greater distance than the latter in their aphelion,
or further part of their orbits;—this singularity is owing to the great
eccentricity of the orbit of Olbers, which is equal to one-fourth part
of its mean distance, while that of Piazzi is but about the twenty-
eighth part of its mean distance from the sun. The other planet was
discovered, September 1, 1804, by Mr. Hardinge, astronomer, at the
Observatory at Lilienthal, near Bremen, in Germany It appears
very small, like a telescopic star of the eighth magnitude. Subsequent observations have determined some of its phenomena. Its
period is four years four months; the inclination of its orbit between
$13°$ and $21°$. Its mean distance from the sun three hundred millions of miles.

The periodical time of Olbers is found to be four years seven
months ten days: and that of Piazzi but little different. The sizes
of these planets are variously stated by different astronomers. Taking
the apparent diameter at a second and a half, the real diameter may
be about one-seventh of that of our earth, or one-half that of the
moon. From Dr. Herschel's observations they appear to be much

smaller; namely, the diameter of Piazzi about one hundred and sixty-two miles, and that of Olbers only ninety-five miles. He also considers them of a different species from the other known planets, and calls them asteroids; as in the clearness of their light they resemble the other planets and stars, while in their size and motion they resemble the comets.

Some of the planets have satellites, or moons, belonging to them; performing revolutions round a centre between themselves and their primaries, and also revolving with them round the sun. Our earth has one moon; Jupiter, four; Saturn, seven; and the Georgium Sidus six, already discovered. From the benefits derived from the influences of the moon on our earth, we naturally infer, that the satellites of the other planets perform the same essential and salutary offices to the respective worlds connected with them.

The means by which mathematicians and astronomers have arrived at an accurate knowledge on this subject, in respect to objects so far removed from their actual reach, are satisfactorily and sufficiently explained in my former lectures.

To afford a rational solution of the globes, and the problems to be performed by them, I shall state the circumstances which confirm us in the belief of the sun being the central body of our system; and of the planets, and their moons, or satellites, shining only by reflecting the light of the sun: also, show how the rotation of these bodies on their axes is ascertained, and explain the cause of eclipses. It is evident to our senses that the earth and the heavenly bodies move round each other. The revolving bodies move in an unresisting medium, on which account these motions are continual, and always regular. It is impossible, by the senses alone, to ascertain which of these has the quickest, and which the slowest motion; or which moves exterior, and which interior, in respect to another ; because the atmosphere revolving with the earth, renders that motion insensible to creatures on its surface ; for the earth has no motion independent of its atmosphere.   Hence, as our senses are insufficient to determine the fact, we must call in the aid of our judgment, which may be confirmed by reasoning on known truths.   Notwithstanding the possibility of the earth moving round the sun, yet as we cannot perceive that it does so by sensible effects, to establish that fact, we will compare this circumstance with effects perceived in familiar instances, and confirmed by undeviating laws.

Suppose a large ball placed on one extremity of a stick, and a smaller one on its other extremity ; to place the whole in such a manner that we may give it a revolving motion round the centre of gravity between the balls, we must duly balance the two balls, by placing their centre of gravity on a point, or pivot.   It is evident to reason, that the centre of gravity of these two unequal bodies must be nearer to the larger, than to the smaller one.   It is a known law in motion, that revolving bodies connected together by an intervening agent, as the sun and planets really are by attraction, must move round a centre ; and that centre be nearer to the larger than to the smaller bodies.   This subject is fully illustrated in my Astronomy ; and the fact of the planets revolving round the sun is now accredited on the best authority, produced from the observations and calculations of the most eminent astronomers and mathematicians.

The sun, from his magnitude, balances all the bodies which circulate round him ; and these bodies are all connected with that luminary by the power of gravity.   It must be obvious to you, that the sun, being the largest of the bodies constituting the solar system, must perform his revolution nearer to the common centre of gravity than any of the others.

Let fig. 2, pl. 32, represent a stick, having a large and a small ball on its extremities, and revolving round a centre at A. The larger ball, revolving nearer to that centre, describes the smaller circle, B C; while the less ball describes the larger circle, D E.

The centre of gravity between the sun and all the planets is not more than the sun's semi-diameter from itself. Thus, by familiar observations, and easy inferences, we are able to establish the sublime and important fact—that the sun is the central body of our system.

That the planets shine only by reflecting the light of the sun, is evident in the effects perceived of the inferior planets *, Venus and Mercury; and also of the moon that accompanies our earth, and the satellites of Jupiter; which never appear bright but when so situated that they receive the sun's rays. In fig. 1, pl. 31, v represents Venus, and T our earth; in this situation of these planets, the side of Venus enlightened by the sun being turned from our earth, that planet does not appear bright, but like a dark spot passing over the sun's disk: when Venus is at c in her orbit, we see part of her illuminated face, and she appears like a half-moon; when removed to A, she appears almost round. Let T, fig. 3, pl. 31, represent our earth, and A B C D the moon revolving round it. First, suppose the moon to be at new, or at A, situated between the sun and the earth; she is not then seen by us, as the whole of her illuminated face is turned from the earth. The moon at the quadrature, or when at B or D, appears like a crescent: when at c, or at full, we see the whole of her illuminated face, because she is in opposition to the sun.

---

* Those planets which revolve in orbits within the orbit of the earth, are called inferior, and those that revolve exterior to the earth's orbit, superior.

In fig 1, pl. 30, the orbits of the planets are represented as round; but, in fact, these bodies perform their annual revolutions in elliptic orbits; the cause of their doing so, I have clearly demonstrated in my Astronomy.

The moon performs her revolution round the centre of gravity between herself and the earth, in a plane inclined to the earth's orbit.

Eclipses do not happen unless the moon be situated immediately, or nearly, at *a*, or *b*, which is the line of her nodes, or central situation in respect to the earth and sun, at the times of new and full moon. When she is at *c* or *d* in her orbit, an eclipse does not take place. If she happen to be at *e*, within a short distance of *a* or *b*, the two nodes, a partial eclipse may take place, and it will be in proportion to the moon's nearness to those points. In all other situations in her orbit, she is either higher or lower than the sun: hence it is, that we do not have eclipses twice every month. When the moon is in that part of her orbit in a line from our globe to the centre of the sun, she intercepts some of his rays in their approach to the earth: when situated in her orbit as at *c*, and near the place of her nodes, the earth casting a shadow behind it, the moon in passing through that shadow is eclipsed. The time a planet is revolving round the sun is called its year; and that in which it completes its rotation round its axis is termed its diurnal motion, because it settles the duration of day and night. All the bodies composing the solar system being globular, and having a progressive motion, naturally turn round a line within themselves. This line is called their axis, round which they revolve while pursuing their progressive range.

That the planets are globular bodies actual observations have determined. As thus: by the aid of glasses we are able to discern spots on some of these bodies; and the different appearances of the spots at different times have been such as must arise from viewing them on the surface of a globular revolving body; namely, their appearing broader when in the centre, in respect to our sight, than when approaching to that central situation, or near to the sides of the revolving body. These effects are most evident on the sun and moon; by which observations the time of the rotation of each of these bodies on its axis is determined   The circular figure cast by our earth in its shadow, as proved at the time of an eclipse of the moon, and that of the moon at the time of an eclipse of the sun, have indisputably established this fact. The fixed stars, as they are called, from their being stationary in respect to our observations of them, always appearing at nearly the same points in the heavens relatively to each other, furnish us, by their apparent diurnal revolution (which is produced by the real motion of our globe on its axis), with the time of the entire rotation of the earth; for by the observation of the situation of a certain fixed star on one evening, and its return to the same spot the next this is ascertained. The sun is so distant from us, that he appears stationary; yet even his motion can be estimated, by our observations of certain spots on his surface.

Having endeavoured to establish the facts of the sun being the centre of the system, and the planets shining only by his light, I shall proceed with the subject of motion.

All bodies revolving on an axis exert a force from their centre, which force is increased in proportion to the greater distance of any part of such bodies from that centre. These effects of the centrifugal force being applicable to the nature and configuration of our earth, we infer that it is greater at the equator than at the poles: and hence it is, that the equatorial part of our earth is larger than any other; which postulatum has been established by actual mensuration and the law of pendulums. The application of the latter to ascertain this circumstance, arose from the motion of this instrument being accelerated by an increased force of gravity, and retarded by its weaker impressions. Hence we infer, that as a pendulum vibrates slower at the equator, that part must have its gravity counteracted by some power, which power is found to be the centrifugal force; this counteracts in a degree the effects of gravity at the equator, and also enlarges that part of the surface of our globe.

In recurring to the other circumstances of the solar system, it becomes necessary to mention certain bodies that are perceived by us at irregular intervals, called comets; but of which no positive theory is established: for neither their periods nor distances are actually ascertained; though calculations have been made of the length of the orbits of some of them, and the time of their revolutions, by observations taken of the velocity with which these bodies move in certain parts of their orbits However, we may suppose the orbits of comets to be very long ellipses, because these bodies are sometimes far beyond our sight, and at others approach very near to the sun, moving with great velocity in their nearest approach to that luminary. From the known laws of motion, and of centrifugal and centripetal forces, we know that all the planets revolve in elliptical orbits, and must therefore be sometimes nearer to the sun, and at other periods further removed from their grand vivifying principle. This change of distance we perceive in respect to our earth, for in winter the sun subtends a larger angle with it than in summer; accordingly, the sun must be nearer to us in the former season than in the latter: but this difference is so small, compared with his absolute distance from us, even in our nearest approach to his splendid animating orb, that no diminution or augmentation of either heat or light is perceived, in consequence of this change of distance.

Let us for a moment leave the small part of the universe to which we belong, and extend our view within the confines of the ethereal expanse; where suns innumerable resplendent shine, animating other planetary worlds that circulate round them. This idea is too grand for our circumscribed comprehensions to appreciate; but the fact is established by the evidences of our senses, and confessedly manifested to us by our reason, which perceives and judges of one thing by another. God has created nothing in vain; and these beautiful luminaries appear like our sun: therefore we naturally infer, that they are suns like that which animates our system, and created for the same wise and beneficial purposes.

Figs. 1 and 2, pl. 34, represent the northern and southern celestial hemispheres; on which are delineated only the stars of the first four magnitudes in each constellation, being all that, during the clearest weather, are evident to our unassisted sight; the remaining classes being only observable by the aid of telescopes. To distinguish particularly such of the stars as are grouped in constellations, appears also the most useful and satisfactory representation of the heavens; for it will serve to facilitate the knowledge of these divisions, by rendering the individual survey of them more distinct.

The stars appear of various sizes to us; but whether this arises from any real difference of size in them, or only from their being situated more or less remote from our earth, we cannot determine:

for we have no means of ascertaining their distances from our globe, not being able to form an angle with any of them ; yet we have the best reason to believe that they are placed at different distances from it. The planets being sometimes nearer to the sun than at others, their orbits must be elliptical; for the centripetal force, or attraction of the sun, acts with greater or less power on them, as they are nearer to that luminary, or further removed from it. Were these bodies constantly acted on by two equal forces; or did the centripetal, or that force which draws them towards the sun, exactly balance the centrifugal, or the force that impels them from that centre ; these bodies would revolve in a circle : but Providence has so ordained, that these circulating worlds should be at different distances from the sun at different periods, by causing sometimes the centripetal force to be greater and sometimes less than the centrifugal ; and hence it is the planets vary in their distances from the sun in different parts of their orbits.

How fitly formed—how duly balanced, is this wonderous system ! Each planet has its appointed station and direction, and implicitly obeys the laws prescribed by God Omnipotent !

> Endless the wonders of creating power
> On earth; but chief on high: through heav'n display'd
> There shines the full magnificence
> Of Majesty divine ! Refulgent there
> Ten thousand suns blaze forth, with each his train
> Of worlds dependent, all beneath the eye
> And equal rule of one eternal Lord.   MALLET.

'Tis true, the mind is lost in the magnificent survey of innumerable worlds on worlds, impelled by Divine command, and revolving in the bosom of immensity ; but surely these effects are not more admirable than the other evidences of the amazing power of the Deity previously contemplated in these lectures, which altogether surpass the utmost stretch of human capacity wholly to develope. Yet the grand survey of the universe, taken in its aggregate magnificence, certainly imparts the most elevating thoughts, displays the profoundest evidences, and affords the most sublimely glorious spectacle of creating Wisdom!!!

In this outline of astronomy, I shall treat more particularly of the circumstances relating to the planet we inhabit; describing such appearances and affections of the heavenly bodies as are perceived and felt by the inhabitants of this earth, and illustrating these by problems subjoined. But for the phenomena of the other circulating bodies of our system, I refer my readers to my former publication on astronomy, where they will find them minutely and scientifically considered; and I trust as clearly elucidated as the sublimity of the subjects will allow.

Having in the preceding lectures explained the nature of the elements belonging to our earth, sufficiently for general and particular application of this knowledge to understanding the variety of climates, &c. that characterise particular places on its surface; and as it is impossible within the limits of this lecture to describe minutely all the various circumstances and effects arising therefrom in its different regions; I shall content myself with elucidating such effects only as are connected with the illustrative problems subjoined.

Fig. 1, pl. 32, represents the globe of our earth, on which are delineated the different continents, countries, oceans, seas and rivers on its surface. This globe, like our earth, revolves on an axis, and can be so placed as to represent the inclination of that axis to the plane of its orbit; which circumstance causes the sun's rays to fall differently on different portions of the earth, and also at the same place at different times; producing the grateful vicissitudes of seasons, and the succession of day and night, to its inhabitants.

The variations in the temperature of the atmosphere, perceived at different situations on the earth, led to the division of it into zones or belts. A B is the torrid or burning zone; B c the temperate, and c D the frigid or frozen zone; each deriving its name from the degrees of heat and cold felt at its respective portion of the earth; which circumstances will be explained in the problems.

The line G H represents that part of our globecalled the equator, being equi-distant from the two extremities of the earth's axis, N, S, called the poles; the one north, the other south of that circle. To render the knowledge of the globes more useful, I shall explain the rationale of them in the practical illustration of their effects; and, by an original mode of elucidating celestial and terrestrial phenomena, I hope to make the inferences so clear, that those who wish for really scientific information, may be able to work the problems, not merely with mechanical facility, which is sometimes the case; but with rational satisfaction. The inefficacy of the former mode of study appears very evidently in those who have been so taught, either this science or any other; for when the habit ceases, the mechanical knowledge vanishes also. This conviction induces me to exercise the reasoning faculties of my pupils in every branch of learning; which method not only affords them rational pleasure in the prosecution of their studies, but also renders the knowledge derived from each branch of instruction—permanently advantageous to them.

The varieties in the soil, climates, and the elementary parts which characterise our globe, are perfectly adapted to the necessities of animal, vegetable and mineral natures, in their different constitutions and species: of this, natural history furnishes the most striking instances, replete with evidences of the wisdom and benevolence of the great Creator!

Whether we examine the minutest works of creating Power, or soar into the regions of expanded ether, all things emit the purest rays of Divine Intelligence! Did then the wise beneficent Creator of all the wonders we contemplate mean we should behold them without understanding the lesson they impart? Certainly not. He meant that the excellency of his works, made evident to our senses and comprehension, should be understood, and duly appreciated. Then surely to pass them unheeded by, must bespeak either gross ignorance, or want of grace in his creatures; for

> The elements and seasons all declare
> For what th' eternal Maker has ordain'd
> The powers of man: we feel within ourselves
> His energy divine: he tells the heart
> He meant, he made us to behold and love
> What he beholds and loves—the general orb
> Of life and being; to be great like him,
> Beneficent and active.   AKENSIDE.

Endless is the theme of universal love; for infinite is the scheme of Providence, unconfined by human laws, by human conception! The small portion of the works of Divine wisdom and beneficence, the perfection of which is immediately within our view, strikes us with wonder, love and awe!—A perfection so complete, so surpassing human reason, that to attempt to understand all its energies would destroy the limited powers of created man: therefore those things that we cannot appreciate, we must admire at a due distance; conceiving, from what we do see, the glories that for wise ends are now hidden from our sight!

CONCLUDING ADDRESS TO MY PUPILS.

IN these lectures, I have had the delightful satisfaction to delineate some of the most striking evidences of the power, wisdom and benevolence of the great Creator! These discoverable attributes of the Deity cannot fail of confirming you in the love and fear of a Being, so wise, so good—of power unlimited—of superintendency universal; and to whom all our thoughts, words and actions are continually present: who guides, by his counsels, the good and humble, who trust in him alone, not relying on their own understanding or virtue; to these he is a shield and buckler, enabling them to resist all assaults of vice and irreligion.

To seek for evidences of the Deity in the operations of his power, and the arrangement of his plans, is no impeachment of our faith; for we may naturally suppose these things were designed for our investigation—in order to confirm us in faith and goodness. The nature of spiritual existence alone is kept from our view and scrutiny, for the salutary purpose of exercising our faith. All investigations of that subject, not included in the Scriptures, must be uncertain speculation; as we cannot compare human affairs and circumstances with things purely spiritual. Enough is revealed to us by the Scriptures, and confirmed by the observation of natural things, for our comfort in this life, and the perfection of our happiness in the state of perpetual existence!

The exercise of our understanding on the subjects of nature, cannot be displeasing to our Creator; but, on the contrary, we may humbly hope, will be acceptable in his sight, when pursued for valuable purposes, and under certain restrictions. Those who presumptuously expect to fathom the depth of Omnipotence, with the limited line of human reason, are sure to be lost in the attempt: evidences of which truth frequently present themselves; for very often we perceive the gift of understanding, bestowed by God, and limited by his wisdom, totally lost in the abyss of thought, never to rise again. But a due employment of our reasoning faculties in investigating the wonders of creation, was certainly designed by the Divine Creator; who meant to enforce the love and practice of virtue, and the observance of order in the arrangement of our affairs, by his blessed example.

As probably this is the last permanent token of my affection I may ever give you; influenced by principle, I shall endeavour to enforce the practice of the indispensable duties of religion and the graces of morality, in order to render you both happy and respectable in this life, and to secure your future everlasting felicity.

The first and most important duties are those that pure religion inspires, and the Scriptures forcibly inculcate both by examples and precepts:—the love and fear of God;—strict observance of his divine commandments;—with faith in the doctrine of our blessed Saviour— that divine doctrine, which so perfectly accords with the known attributes of the Deity, and breathes nothing but love, peace and joy. True religion is not depressing and gloomy; but, on the contrary, induces cheerfulness by the consoling and enlivening ideas it imparts. A sense of religion should be expressed more in conduct than in words. Suspect those who venture on all occasions to call on the name of God; which familiar use of that word of sacred import, derogates from the respect with which it should ever be uttered.

In public and private worship, address the Deity with humility and devotion; not suffering the objects of sense, which in the former situation surround you, to draw your attention from that reverence with which you should ever address your Maker. A sense of the power of the Deity, and the known insufficiency of our nature to perform what is right without his divine assistance, should lead us, in the next place, on *every occurrence* of human life, silently—not Pharisaically—to address ourselves to God; to enable us to act as seemeth best in his sight—for very important results are frequently produced by apparently slight occasions.

I will not attempt to expatiate further on the doctrines of religion, as these discussions more properly belong to such excellent divines as have been called to this sacred task; and whose wisdom, virtue and intelligence adorn the doctrines they inculcate. Thus duly qualified, appointed and invested, they can, with more propriety, also enforce the precepts of the gospel, and do justice to the holy cause of religion.

The moral duties next present themselves, and these I shall consider in every relation of female life—as children, sisters, friends, wives, mothers and associates; and shall endeavour to furnish you with such rules of conduct, as, if strictly observed, cannot fail of rendering you respectable and desirable in each of these respective affinities.

The grand foundation of all these must consist of the permanent materials arising from a due cultivation of the christian graces of justice, prudence, temperance and fortitude. Be just in your thoughts, words and actions;—prudent in the arrangement of your affairs;—temperate in your wishes and enjoyments;—patient in bearing injuries, and resigned under all the dispensations of Providence. Duly impressed with the importance of these cardinal virtues, and diligently observing their dictates; let your situation in this life be exalted or otherwise, or however subjected to vicissitudes; you will preserve that serenity of mind—that vigour in performance —and that dignity of character, which ever result from principle; and can never be depressed, or rendered torpid or abject, by the utmost depression of situation or circumstances.

As children, be obedient to the dictates of your parents; grateful for all their exertions for your benefit and happiness; and affectionately attentive to all their wants and desires. As friends, be faithful and reasonable; not selfishly wishing to lessen extended and general influences of friendship, by depriving others of the attentions of your friend, who are entitled, either by consanguinity or correspondent regard, to a share of the affections so necessary perhaps to their happiness, as well as to your own. As sisters, be affectionate; and endeavour by every good office to exhibit that generous interest, which regards the welfare, respectability and happiness of those to whom you are so nearly allied. When wives, consider the solemn oath pledged before God, and strictly obey its mandates. Let cheerful acquiescence evince your affection towards your husband. Be the softener of his cares—the sympathizer in all his anxieties; and should unforeseen misfortunes overtake him, then will be the time to show him the strength of your understanding, the purity of your mind, and the nature of your affection. Excite his fortitude by your example—

lessen his anxiety by your vigorous resistance of calamity—and diminish the pressure of misfortune by your active exertions. This will be the season for more particularly displaying the moral graces of justice, prudence, temperance and fortitude. As mothers, remember you once were young. Let your experience and mature judgment direct and admonish your children; but let your admonitions, restraints and directions be softened by maternal affection. If the case require corrosives, though these may be salutary in some cases, use them like a skilful surgeon, firmly, not timidly; and do not fail to prepare the healing balm—let the affection which dictates the measure render it supportable: this will soften the necessary inflictions of the sharpest reproof, and doubtless effect a cure; whereas the wounded feelings, when left to the impressions of correction only, may become callous and incurable. Let not a mistaken fondness, and desire to make your children happy, induce you to allow them indulgences which are pernicious, either in their nature or consequences: for remember—children are not bestowed for the indulgence of affection only, but demand your most vigilant care of their health, morals and religious principles.

It now remains for me to say something of the duties required of you in society; and first, in addition to the fundamental ones commanded by God himself, and included in the cardinal virtues, you should render all possible service to your fellow-creatures—by protecting, instructing and admonishing them; soothing their sorrows, and charitably relieving their distresses, according to their several claims on your christian character.

In society, be unassuming, obliging, charitable; let your benevolence be as conspicuous in judging of conduct, as in bestowing the gifts of abundance. Cultivate a cheerful disposition, and impart its emanations; but let your gaiety be tempered by sedate thought and reflection. Be not anxious about the domestic affairs of others: curiosity is trifling and impertinent, unless excited by the laudable motive of contributing, by our counsel or assistance, to the comfort and happiness of our fellow-creatures. Avoid gossiping or talking of other people's affairs; for this practice bespeaks a weak and vacant mind, and derogates from the modesty, delicacy and refinement of the female character.

Let humility, urbanity and magnanimity adorn your exterior. Suffer not the little infelicities of domestic arrangement to enfeeble your mind—be great in thought, word and deed. In mixed society, avoid that littleness of mind that attends more to external circumstances than to interior worth. Let your duty to God and man, in every connection of your life, and a due cultivation of your reason, pre-occupy your thoughts; and divert them from the fallacious allurements and inconsistencies of folly, and the irrational preponderance of prejudice and fashion. Avoid the vicious, however exalted by rank, or aggrandised by wealth; and respect and distinguish virtue wherever it may appear. Always prefer the society of well-informed and religious persons: and, though I disapprove of particular respect being paid to rank or condition, when unaccompanied by virtuous conduct, yet when those elevated by birth and fortune are also distinguished by merit and religious graces, the laws of society demand that they should receive respect and deference; and this testimony of having received that polish, which understands the forms of polite society, does no more injure internal dignity of mind, than the diamond's polish does its nature; but, on the contrary, like that effect on the valued gem—politeness exhibits more evidently the graces of a virtuous and cultivated mind.

Having shown you the particular intention of the lectures I have had the pleasure of delivering to you, and given you my best advice for your future conduct in life, I cannot conclude this Address better than by adverting to those connections in life, which, being dependent on yourselves, require much consideration, and which I think it my duty to impress on your minds,—the indispensable qualifications of both a friend and a husband—*religious principles and practice :*—never make your choice of either of these till you have discovered that they not only profess to be religious, but are truly so, in thought, word and deed.

When the lips that delivered these maxims are mouldering in the dust, may their respective impressions remain on your hearts ! And should the tear of regret flow on your cheeks, let this reflection be your consolation—that the spirit that dictated them, disrobed of its mortal habiliments, may, through the merits and intercession of a Saviour and Redeemer, be enjoying that exalted felicity which is perfect in its nature—perpetual in its duration.

APPENDIX.

CHARACTERS OF BODIES BELONGING TO THE SOLAR SYSTEM,

AND OF

THE CONSTELLATIONS OF THE ZODIAC.

Planets Characters.	Names of the Planets	Twelve Signs of the Zodiac.		The Letters of the Greek Alphabet, and their Names.			
		Characters.	Names.				
☉	Sun.	♈	Aries the Ram.	A α	Alpha. 1	N ν	Nu.. ..13
☽	Moon.	♉	Taurus . the Bull.	B β ϐ	Beta 2	Ξ ξ	Xi.... 14
⊕	Earth.	♊	Gemini . the Twins.	Γ γ ϝ	Gamma . 3	O o	Omicron 15
☿	Mercury.	♋	Cancer .. the Crab.	Δ δ	Delta ... 4	Π π ϖ	Pi... ..16
♀	Venus.	♌	Leo. the Lion.	E ε	Epsilon.. 5	P ρ ϱ	Rho .. 17
♂	Mars.	♍	Virgo . the Virgin.	Z ζ ζ	Zeta. ..6	Σ σ ς	Sigma ..18
♃	Jupiter.	♎	Libra . the Balance.	H η	Eta.. 7	T τ ϑ	Tau 19
♄	Saturn.	♏	Scorpio.. the Scorpion.	Θ ϑ θ	Theta... 8	Υ υ	Upsilon 20
♅ ⛢ {	Herschel, or Georgium Sidus.	♐ ♑	Sagittary the Archer. Capricorn the Goat.	I ι ι	Iota .... 9 K κ Kappa . 10	Φ φ X χ	Phi.. 21 Chi.. 22
☊	Ascending Node.	♒	Aquarius the Water-Bearer.	Λ λ	Lambda 11	Ψ ψ	Psi .23
☋	Descending Node.	♓	Pisces .. the Fishes.	M μ	Mu.... 12	Ω ω	Omega .24

CHARACTERS OF THE AFFECTIONS OF THE PLANETS AND THEIR SIGNIFICATIONS:

ALSO OTHER CHARACTERS EXPLAINED THAT ARE USED IN THE EPHEMERIS.

Characters.	Names.	Signification.
☌...	Conjunction.	When two planets are in the same sign of the ecliptic.
☍.... ..	Opposition..	When two planets are in opposite signs of the ecliptic.
∗..........	Sextile ...	When two planets are 60 degrees from each other.
□.	Quartile.	When two planets are 90 degrees from each other.
△..	Trine..	When two planets are 120 degrees from each other.

Mark.	Signification.
° ... , ... " ...... ‴	A degree. A minute of a degree. A second of a degree. A third of a degree. } There are 60 minutes in a degree, and 60 seconds in a minute.

A. M. or m. ...... Ante-Meridian, or Morning.
P. M. or a. ..... Post-Meridian, or Afternoon.
h. m. s. . ..... Hours. Minutes. Seconds.

TABLE

OF

SPECIFIC GRAVITIES,

SUPPOSING RAIN WATER 1000.

Refined Gold 19,640

Refined Silver 11,091

Lead 10,130

Copper............. 9,000

Iron.............. 7,645

Tin 7,550

Copper Ore 3,775

Lead Ore 6,800

Adamant, or Diamond . . . 3,400

Cornelian 2,568

Lapis Lazuli 3,054

Lapis Calaminaris...... 5,000

Common Glass 2,620

Crude Mercury 13,593

Mercury distilled 511 times .... } 14,110

DECLINATION

OF

THE MAGNETIC NEEDLE,

OBSERVED IN LONDON AT DIFFERENT TIMES.

Years.	Declination.	
1576	11° 15′	} East.
1580	11 11	
1612	6 10	
1622	6 0	
1633	4 5	
1634	4 5	
1657	0 0	} West
1665	1 22$\frac{1}{2}$	
1666	1 35$\frac{1}{2}$	
1672	2 30	
1683	4 30	
1692	6 0	
1700	8 0	
1717	10 42	
1724	11 45	
1725	11 56	
1730	13 0	
1735	14 16	
1740	15 40	
1745	16 53	
1750	17 54	
1760	19 12	
1765	20 0	
1770	20 35	
1774	21 3	
1775	21 30	
1780	22 10	
1785	22 50	
1790	23 34	
1795	23 52	
1800	24 7	

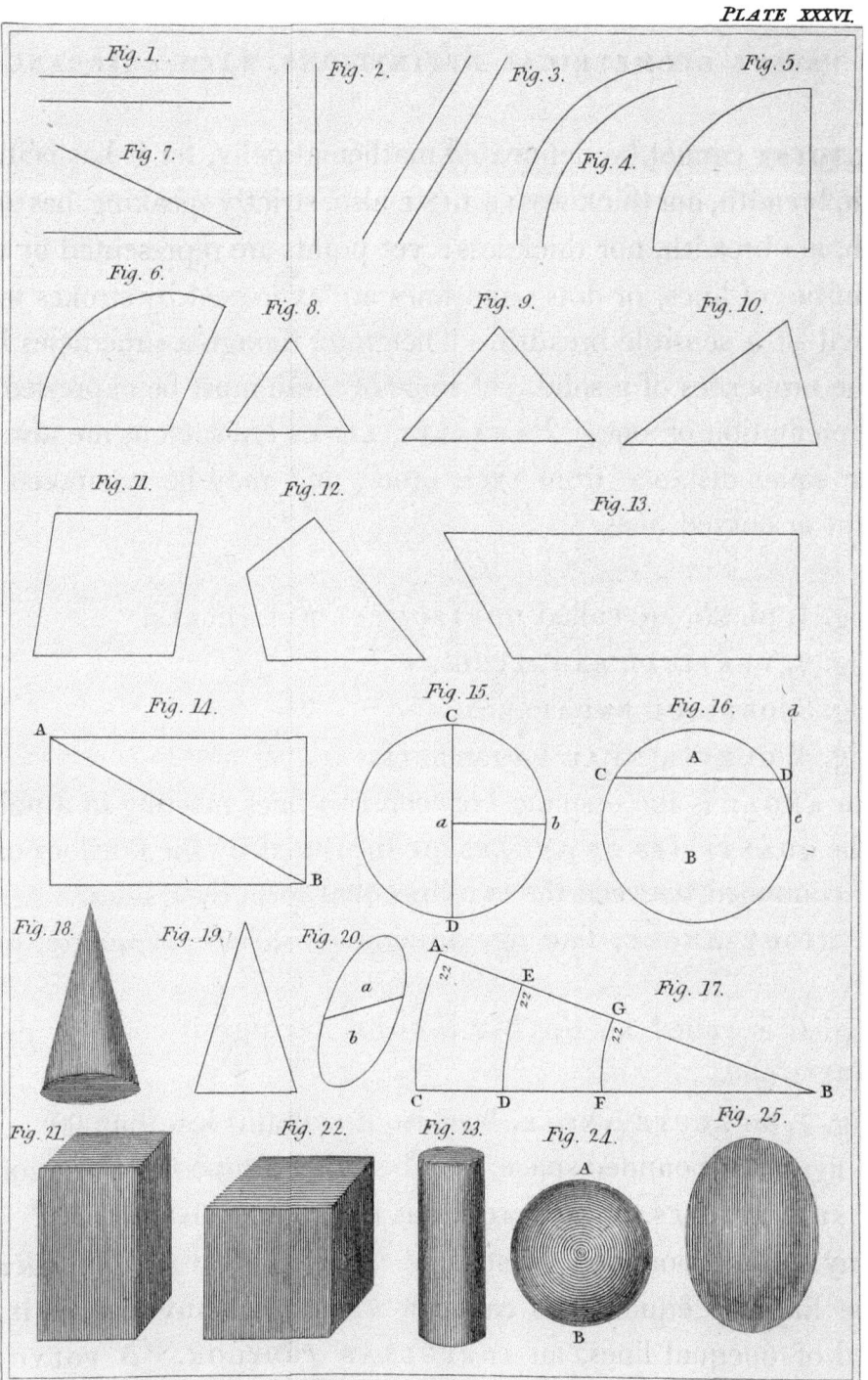

PLATE XXXVI.

M.Bryan del.

H.Mutlow sculp.

SOME USEFUL GEOMETRICAL DEFINITIONS, WITH REFERENCES.

A POINT cannot be delineated mathematically, for it has neither length, breadth, nor thickness; a LINE also, strictly speaking, has only length, not breadth, nor thickness: yet points are represented by the extremities of lines, or dots; and lines are expressed by strokes with a pencil of a sensible breadth. Therefore, though a superficies has not the properties of a solid, yet some of these must be expressed in a representation of one. PARALLEL LINES are such as are always at the same distance from each other, and may be composed of straight or curved lines.

Fig. 1, pl. 36, are called HORIZONTAL PARALLELS.

Fig. 2, VERTICAL PARALLELS.

Fig. 3, OBLIQUE PARALLELS.

Fig. 4, CURVILINEAR PARALLELS.

An ANGLE is the opening between two lines meeting in a point.

The QUANTITIES OF ANGLES are measured by the portions of a circle contained between the two lines that form their sides.

A RIGHT ANGLE, like fig. 5, contains 90°, or a quarter of a circle.

Fig. 6 is called an OBTUSE ANGLE, because it contains more than 90°; and

Fig. 7, an ACUTE ANGLE, because it contains less than 90°.

A figure, or bounded space, may be either a superficies or a solid.

A SUPERFICIES, or SURFACE, has length and breadth only

Any surface bounded by straight lines is called a POLYGON: if all the lines be equal, it is called a REGULAR POLYGON; if it be formed of unequal lines, an IRREGULAR POLYGON. A POLYGON has always as many sides as angles. The figure is sometimes expressed by the number of its sides, and sometimes by its number of angles. A POLYGON of three sides is called a TRIANGLE; of four sides, a QUADRANGLE; of five sides, a PENTAGON; of six sides, a HEXAGON; of seven sides, a HEPTAGON; of eight sides, an OCTAGON; of nine sides, a NONAGON; of ten sides, a DECAGON; of eleven sides, an UNDECAGON; of twelve sides, a DUODECAGON. To express a superficies with a greater number of sides, we say, a polygon of thirteen or fourteen sides, &c.

TRIANGLES are expressed according to the relative length of their sides.

Fig. 8 is called an EQUILATERAL TRIANGLE, having all its sides equal.

Fig. 9 an ISOSCELES TRIANGLE, having only two equal sides.

Fig. 10, a SCALENE TRIANGLE, having all its sides unequal.

A figure of four sides, which has two pairs of sides parallel to each other, is called a PARALLELOGRAM.

When the sides of a regular parallelogram has all its sides equal, but not all at right angles to each other, the figure is called a RHOMBUS, like fig. 11. When all the sides of a parallelogram are oblique to each other, and only the two opposite ones equal, it is called a RHOMBOID.

When a quadrangular figure of four sides has none of them parallel, like fig. 12, it is called a TRAPEZIUM.

If it have two sides only equal, like fig. 13, it is called a TRAPEZOID.

A DIAGONAL is a right line drawn between two angles from their two points, like A B in fig. 14.

AREA means the space contained between boundary lines.

A CIRCLE is a curved line, every-where at an equal distance from the centre, like fig. 15.

The RADIUS of a circle is the line drawn from its centre to its circumference, like *a b*, fig. 15: this is also the semi-diameter of the circle.

The DIAMETER of a circle is the line drawn across its area from two opposite points in its circumference, like C D, fig. 15.

A SEGMENT of a circle is a portion of a circle cut off by a straight line, like A and B in fig. 16. The line that cuts off the segment of a circle is called the CHORD line, represented by C D, fig. 16.

A TANGENT is a line that just touches a circle, like *d c* in fig. 16.

To measure the quantity of an angle, we place one point of the compasses on the angular point, as at B, and describe an arc of a circle, which will show the quantity of the angle. All circles, large and small, contain 360°. It is not the length, but the opening of the lines that determines the quantity of an angle, as may be easily understood by the equal portions of the larger and smaller circles included between the lines A, C, D, E, and F G, in fig. 17, each containing 22° of their respective circles.

A SOLID has length, breadth and thickness.

A CONE is a solid contained within a convex surface, diminishing from a circular base; as represented fig. 18. The upper part of a cone is called its VERTEX: if a cone be cut through the centre, from its vertex to the base, the section will be a triangle, like fig. 19.

If a cone be cut obliquely to its axis, the section will be an ELLIPSIS, like fig. 20. If the cone be cut obliquely from its side parallel to its opposite side, the section will be a PARABOLA, like *a* and *b* in fig. 20.

The sides of a prism are PARALLELOGRAMS; and the ends, or base, POLYGONS parallel to each other.

PRISMS are expressed according to the form of their bases.

If the base of a prism be a parallelogram, it is called a PARALLELOPIPEDON; which is represented by fig. 21.

If all the sides are square, the solid is called a CUBE, like fig. 22.

A PYRAMID is contained within a number of triangular planes. The pyramid is also expressed by the figure which forms its base, as a triangle, &c. &c.

A CYLINDER is a solid body, like fig. 23, rising from a circular base; and when cut obliquely to its axis, the section is an ellipsis.

A cone, prism and cylinder, may either be upright or oblique.

A SPHERE is a solid, having every part of its superficies convex.

The line A B, round which a sphere revolves, is called its axis. The terminations of this line at A B are called the poles.

Every section of a sphere is a circle; and every section taken through the centre of a sphere is called a great circle, and divides the globe into two hemispheres. Every portion of a sphere cut off is called a segment.

A SPHEROID is a solid of an oval form, like fig. 24.

EXPLANATION OF CERTAIN SIGNS USED TO EXPRESS ARITHME-
TICAL AND GEOMETRICAL PROPORTIONS OR QUANTITIES.

SIGNS.	EXPLANATIONS.
°	Is used to express degrees, as 23°.
′ ″	Minutes of a degree, 23° 30′. Seconds of a degree, 23° 30′ 2″.
+ Plus or more	Used to express addition.
= denotes equality	These may be thus employed $5 + 2 = 7$, that is 5 added to 2 equal to 7.
− Minus or less	Denoting Subtraction $9 - 4 = 5$, that is, 9 lessened by 4 is equal to 5.
× Multiplication	$6 \times 8 = 48$, that is to say, 6 multiplied by 8 is equal to 48.
÷ Division	$12 \div 4 = 3$, namely, 12 divided by 4 is equal to 3.
: :	$3 : 6 :: 8 : 16$, that is, as 3 is to 6 so is 8 to 16.
$12 - 4 + 2 = 10$	Shows that the difference between 12 and 4 when added to 2 is equal to 10.
√ Square Root	This mark prefixed to a number, shows that the square root of that number is to be extracted.
∛ Cube Root	Signifies that the cube root of the number placed after it is to be extracted.
a^2	This sign placed to a number, shows that number must be squared, or raised to the second power ; that is, multiplied by itself.
a^3	Shows the number must be cubed, or raised to the third power ; that is, multiplied twice by itself.
Q. E. D.	Signifies, which was to be demonstrated.
Q. E. I.	Signifies, which was to be found.
Q. E. F.	Signifies, which was to be done.

THE STARS VERY CONSPICUOUS AT THE LATITUDE OF LONDON,
WITH THE NAMES OF ALL THE CONSTELLATIONS VISIBLE
THERE, EVEN THOSE IN WHICH THERE ARE NO REMARKABLE
STARS.

URSA MAJOR has seven very conspicuous stars, six of the second
magnitude, and one of the third magnitude; which latter is at the
extremity of his tail, and is called *Benetnash*. The two nearest to
his head are called the *Pointers;* because a line drawn through them
points to the north pole. The *Pointer* in his shoulder is called
Dubhe; the star on his tail nearest his head is called *Astroth;* the
next to it on his tail *Mizar*.

URSA MINOR has seven stars situated like the seven of ursa
major; but they are not so conspicuous, being of the third and fourth
magnitudes. That at the extremity of his tail is called the *Pole Star:*
and the one on his shoulder *Hoehab*, distant about 2° from the north
pole.

DRACO THE DRAGON has four very conspicuous stars of the
second magnitude: the one on his head is called *Rasteben*.

CEPHEUS KING OF ETHOPIA has three stars of the third magnitude: the one in his shoulder is called *Alderamin*.

CYGNUS THE SWAN has one star of the second magnitude,
called *Deneb;* and one of the third magnitude, called *Albireo*.

LYRA THE HARP has one of the first magnitude, called *Wega*.

HERCULES has one star of the second magnitude, called *Ras
Algethi*.

In CORONA BOREALIS, or the NORTHERN CROWN, there is one
star of the second magnitude, called *Alphecca*.

BOOTES THE KEEPER OF THE GREAT BEAR has one star of
the first magnitude, called *Arcturus;* one of the third magnitude on

his belt, called *Mirac;* one of the second magnitude in the heart. There are two conspicuous stars in his dogs, which are called *Asterion* and *Chara.*

Neither LEO MINOR, THE LYNX, THE CAMEL, nor THE REIN DEER, has any conspicuous stars.

AURIGA THE WAGGONER has one star of the first magnitude in his left shoulder, called *Capella,* and a star of the second magnitude on his right shoulder.

PERSEUS has one star of the second magnitude at his breast, called *Algenib;* and one of the second magnitude in *Medusa's Head,* called *Algol.*

CASSIOPEIA has one star of the second magnitude on her breast, which is called *Schodir;* and one of the third magnitude.

ANDROMEDA has three stars of the second magnitude: one of which is in her head; one in her girdle, called *Mirach;* one at her heel, called *Alnmak.*

PEGASUS THE WINGED HORSE, and EQULEUS THE COLT. *Pegasus* has three stars of the second magnitude: the one of which on his leg is called *Scheat:* and two on his wing; one called *Markab;* the one below the other, *Algenib.*

DELPHINUS THE DOLPHIN has five stars of the third magnitude.

In VULPECULA, and ANSER THE FOX AND GOOSE, and in SAGITTA THE ARROW, there are no conspicuous stars, for they are all less than the third magnitude.

In AQUILA THE EAGLE there is one star of the first magnitude, called *Athair.*

In THE BULL OF PONIATOWSKI, no very conspicuous stars.

In SERPENTARIUS THE SERPENT BEARER, one star of the second magnitude in his head, called *Ras Alhague.*

In THE SERPENT, one star of the second magnitude.

In COMA BERENICE, no very conspicuous stars.

In VIRGO, one star of the first magnitude in her head, called *Spica.*

In LIBRA are two stars of the second magnitude, one in each scale.

SCORPIO has one star of the first magnitude, *Antarus*, in her heart, and one of the second in her head.

In SAGITTARIUS there are no very conspicuous stars.

In CAPRICORNUS there are no conspicuous stars.

AQUARIUS has no conspicuous stars.

In PISCES there are no conspicuous stars.

In ARIES there is one star of the second magnitude in his forehead.

In TAURUS, one star of the first magnitude in his eye, called *Alderbaran;* and one of the second magnitude in his north horn.

In GEMINI, one star of the first magnitude, *Castor*, in the cheek of *Castor;* and two of the second magnitude, one on the neck of *Pollux*, and one on his foot.

CANCER has no conspicuous stars.

LEO has one star of the first magnitude in his heart, called *Regulus;* and one of the first magnitude in his tail, called *Donobella;* and two of the second magnitude in his back.

CRATER THE CUP, which is under *Leo's* foot, has no conspicuous stars.

SEXTUS THE SEXTANT has no conspicuous stars.

HYDRA has one star of the second magnitude in his heart, called *Alphard.*

In CANIS MINOR there is one star of the second magnitude, called *Procyon.*

In THE UNICORN there are no conspicuous stars.

In ORION there is one star of the first magnitude in his right shoulder, called *Betelguse;* one of the first magnitude in his right shoulder, called *Rigel;* one of the second magnitude in his left shoulder, called *Bellatrix;* three of the second magnitude in his girdle.

In ERIDANUS THE RIVER are two stars, one star of the first, and one of the second magnitude.

In CETUS THE WHALE is one star of the second magnitude in his jaw, called *Menkar.*

In the SOUTHERN FISH, one star of the first magnitude, *Fomalhaut.*

In LEPUS THE HARE there are no very conspicuous stars.

CANIS MAJOR, to the right of *Lepus,* and below *Orion,* has one star of the first magnitude in his jaw, which is called *Sirius,* and two of the second magnitude.

ARGO, NAVIS, and CORVUS THE SHIP AND CROW, have no very conspicuous stars visible at our latitude.

THE GENERAL PRINCIPLES OF THE CELESTIAL AND TERRESTRIAL GLOBES AND ARMILLARY SPHERE EXPLAINED, PREPARATORY TO THE PROBLEMATICAL ILLUSTRATIONS, REFERABLE TO THEM; AND ALSO TO THE TERRESTRIAL AND CELESTIAL MAPS.

ON a celestial globe, represented by fig. 1, pl. 33, are shown the fixed stars of the heavens which are alternately exhibited to our view by the daily revolution of our earth on its axis.

E F G is called the horizon, because it represents that circle which bounds the sight of a spectator on the earth, when his situation is brought to the zenith or top point of the globe. On the horizon are delineated the twelve signs of the zodiac, through which the sun apparently performs his annual revolution: the days and months of the year are also delineated on the plane of the horizon.

Of the four cardinal points, ғ represents the north, ᴇ the south, ɢ the east, and the supposed opposite point to that is the west point of the horizon.

ᴀ ʙ ᴄ ᴅ exhibits the situation of the meridional circle which encompasses the globe; representing the situation of that supposed circle of the heavens, passing through the north and south points of the compass or horizon, and over our heads through the zenith or highest point, ᴢ, and descending to the nadir, ɴ, or lowest point of the globe immediately opposite to the zenith.

ɪɪ and ɪ exhibit the two extremities of the axis on which the globe revolves, which is supposed to be the continuation of the axis on which the earth performs its diurnal revolution. For as our earth actually revolves round this line within itself, we consequently see all the fixed stars, excepting the one situated at the pole, apparently revolving in a direction contrary to the known revolving direction of our earth. The line ᴋ ʟ, encompassing the globe, represents the circle in the heavens called the equator: this is 90° from each pole. ᴍ ɴ is the ecliptic, or sun's apparent path in the heavens. Mathematicians divide all circles into 360 parts, or degrees: hence the great circles of the globe, namely, the horizon, meridian, equator and ecliptic, are always so divided. The meridian is graduated into four quarters; two of which are counted from the equator towards the poles, for the purpose of finding the latitude of a place, that is, its distance from the equator, north and south; and the other two from the poles towards the equator: so that when we elevate the pole to the given latitude, the correspondent degrees or place comes to the zenith point of the globe.

The equator, ᴋ ʟ, is divided into two parts counted, on English globes, from London, east and west. These are used to ascertain the longitude of places, which is estimated by their distance from those points to 180°. One intersection of the equator and ecliptic is seen in fig. 1, pl. 33, at the eastern edge of the horizon. o p represents a quadrant of altitude, which serves to show the latitude and longitude of the stars when placed at c, over the pole of the ecliptic, and brought down to that circle. R is a dial which shows the time of a celestial appearance or station. It is divided into twenty-four hours, and describes these while the globe revolves once on its axis; thus according with the apparent revolution of the heavens, and agreeably to the received notion of the earth's revolution being performed exactly in that time. The deviation from the space of twenty-four hours in the earth's revolution, is too trifling for general problematical purposes; but it is accurately stated in my Astronomy, and the cause duly explained. s T is the magnet-box, containing a touched needle and compass card, to enable us to set the globe correctly, allowing for the variation of the needle at the given time and place. The meridian passes through the two cardinal points of the horizon called north and south; and the equator through the other two, namely, the east and west.

Our earth revolving round the sun on an inclined axis, sometimes one, and at other times another part of it is in a direct line with him, or nearer and further from that direction, as is exhibited by fig. 2, pl. 31.

No part of our earth, excepting that situated within 47°, called the torrid zone, as from m to b, ever has the sun in a perpendicular or straight line with it. The two circles bd and M o, which are 23° 30′ distant from the equator, are called the tropics; because in our observations of the sun on the meridian, day after day, when he arrives at either of those, he appears to return towards the other.

The circles situated at E·*h* and *g* D, exactly 23° 30′ distant from the poles, are called the polar circles. The regions included within these are excessively cold; the sun's rays falling more obliquely on those parts than on any other portion of the earth.

The space between the tropics is called the torrid region or zone, because the sun's rays fall perpendicularly and abundantly there: and the spaces between the tropics and polar circles are called temperate zones, they being situated between the two extremes. In the northern temperate zone we are placed.

Fig. 1, pl. 32, represents the terrestrial globe, of which I have previously spoken, and acquainted you that on its surface are delineated the different parts of the earth. The horizon, meridian, hour circle, and other appendages of this globe, being the same as those of the celestial one, require no further explanation. The latitudes of places are estimated by their distance from the equator, G H: those situated towards the north pole, N, are said to be in the northern, and those on the south side of the equator in southern, latitude.

The land part of our globe is divided into two grand divisions, as may be seen on the planisphere, figs. 1 and 2, pl. 35; on which are represented the various places situated on the earth. The boundary line of each figure represents the equator; on which the longitude of places is counted. All the circles drawn parallel to the equator on the maps 1 and 2, are parallels of longitude; but the graduated ones are those by which the longitude of places is ascertained; and the lines drawn through these are parallels of latitude. This projection of the earth is the best plane one that can be used for problematical illustration. Fig. 1 is the northern, and fig. 2 the southern hemisphere. The circle of figures exterior to the equator of each map, serves as a clock in problematical references respecting time.

Fig. 2, pl. 33, is the armillary sphere; so called from its consisting only of rings. In this sphere, A B represents the equator; C D the ecliptic, intersecting the former in two points, the same as on the globes. These two points are called the equinoctial, being those in which the equator of our earth is exactly in a line with the centre of the sun; which situation of the sun and earth is called the equinoxes, and happens in March and September. The tropics are represented in fig. 2, pl. 33, by E F; and the polar circles by G H, the meridian by I K, and the horizon by L M.

The armillary sphere has the earth, T, placed naturally, that is, in the centre of the imaginary circles of the heavens. The poles of the earth, N, O, are continued to the heavens, at N O. And each circle on the earth corresponds with the circle of the heavens passing over it. This instrument conveys a distinct idea of each imaginary circle of the heavens, but does not exhibit the stars, and therefore is not very extensively useful in problematical elucidations.

I shall conclude this introduction to the problems with a few remarks applicable to the globes, maps and armillary sphere.

The circles which intersect the poles of the terrestrial globe and planisphere are called circles of latitude, because on these the distances of places from the equator are counted; but these are called circles of declination on the celestial ones. The circles drawn parallel to the equator on the terrestrial globe are called parallels of latitude, and those drawn parallel to the ecliptic of a celestial globe are called the same.

The celestial globe has two pairs of poles, one to the ecliptic, and one to the equator. That to the ecliptic is to show the latitude of the stars counted from the ecliptic; that to the equator, their declination: which latter are the extremities of the axis on which the globe revolves.

On the ecliptic circle the longitude of the stars is reckoned. The distance of a star from the equator is called its declination; and this declination is found on the meridian, which has the degrees numbered from the equator to show the distance of any star from that circle. As the latitude of the stars is counted from the ecliptic, and the degrees of the meridian answer only for their distances from the equator, in order to find the latitude of a star, we use the quadrant of altitude, which has 90° graduated on its surface: so that, placing the ninetieth degree over the pole of the ecliptic, and bringing the quadrant down to that circle, we are able to ascertain the distance of a star from the ecliptic, or its latitude. Eight parallel lines are usually described on a celestial globe, on each side of the ecliptic, to include the latitude of the planets; but the last-discovered ones exceeding that boundary, these parallels do not answer for them, but only for those previously known.

ASTRONOMICAL AND GEOGRAPHICAL PROBLEMS.

PROBLEM I.

To find the Latitude and Longitude of a Place on the Earth.

THE places of the earth are laid down on the surface of the terrestrial globe, so as to answer exactly to their real situations on it.

To find the latitude and longitude of a place, bring the place to the graduated meridian, and observe what degree lies over the place, counted from the equator, this being its latitude. The degree of the equator intersected by the meridian at the same time shows its longitude. If the place be on the north side of the equator, its latitude is northern; if on the south side, southern. If it be east of London, its longitude is eastern; if west, western. Latitude is never more than 90°, being reckoned from the equator to the pole; but longitude is counted 180° east and west.

PROBLEM II.

To find the Latitude and Longitude of a Place on the Planisphere.

Observe what degree of the meridian is intersected by the parallel of latitude on which the place is situated: this will show its latitude; and continuing a straight line to the equatorial circle, the intersected degree will be its longitude.

PROBLEM III.

To find a Place situated in a particular Latitude and Longitude.

Bring the given degree of longitude to the meridian, and observe what place lies under the given degree of latitude.

PROBLEM IV.

To find all the Places that have the same Latitude.

The latitude of places being ascertained by bringing them to the meridian, and observing the degree lying over each place; on turning the globe, all places that pass under the same degree of the meridian have the same latitude with each other.

PROBLEM V.

First, to find the Difference of Latitude between two Places situated on the same Side of the Equator, and secondly, those situated on opposite Sides of it.

When both are on the same side of the equator, bring each place to the meridan, and then subtract the less latitude from the greater, and the remainder will be the difference required. If the places be one north and the other south of the equator, add them together for the whole difference.

PROBLEM VI.

To find all Places having the same Longitude.

As we count the longitude of places on the equator, all the places under the same meridian at one time have the same longitude with each other.

PROBLEM VII.

To find the Difference of Longitude between two Places.

Bring each place to the meridian, and observe their longitude. Then, if both be either east or west, subtract the less from the greater longitude, to find their difference. If one be east and the other west, add the two quantities of their longitude together, to find the whole difference in longitude between them.

PROBLEM VIII.

To find all the Places having the Same Latitude and Longitude on the Planisphere.

All those places that are on the same line drawn from the pole of fig. 1 and 2, pl. 35, and intersecting the same degree of the exterior circle or equator, have the same longitude. Those which are situated on the same parallel to the exterior circle of fig. 1 and 2, or equator, have the same latitude.

PROBLEM IX.

To find the Difference of Longitude and Latitude between Places on Fig. 1 or 2, Pl. 35.

Subtract the less from the greater number on the equator, or boundary line, of fig. 1 or 2, for the difference of longitude between places situated on the same side of the meridian of London; but if one be in eastern and the other western longitude, add the numbers together.

For the difference of latitude between two places, when one is north and the other south, add them together; but when both are north or south, subtract one from the other, as before directed.

PROBLEM X.

To find, by the Terrestrial Globe, the Hours of the Day at Different Places within the Torrid and Temperate Zones.

When the sun is on that line which passes over our heads from the north to the south point of the horizon, at whatever part of the earth we are, it is said to be mid-day, or twelve o'clock; because the sun in that situation, at any place, is at his highest point above the horizon of it for that day.

The earth revolving on its axis from west to east, if we imagine the sun to be, as it really is in respect to us, stationary, of course the places situated east of others must have twelve o'clock the earliest. The equator of the globes and planisphere is divided into twenty-four parts, to show those differences. The twenty-four hours are also engraved on the dial, or hour circle; and the index of the globe revolving with it, if we bring any place to the meridian, and fix the index to twelve o'clock, the time the sun is on the meridian of that place, by turning the globe naturally, or from west to east, we find the difference of time at two places by the dial. If the first observed place be at the meridian, and the second place be east of it, we shall find it is afternoon at the second place when noon at the first. If the second be situated west of the given meridian, it will be before noon at that place. This difference of time is always the same for every day in the year; as it depends only on the direction of the earth's diurnal revolution on its axis, compared with the relative situations of places, which never vary.

PROBLEM XI.

To find the Difference of Time between Places on the Planisphere,
Pl. 35.

The hours are marked on the equator; hence, if we observe the longitude of places, and notice how many hours are between their situations, east or west of each other, we shall ascertain the whole difference of time between them; and supposing it to be twelve o'clock, or any other hour, at one of these places, we shall find what o'clock it is at the other.

PROBLEM XII.

To find the Difference of Time between Places by Calculation only.

The equator contains 360°, or parts, which are collected into

twenty-four houis; hence 15° of the equator, or of the earth's revo--
lution on its axis, are equal to an hour, and each degree is equal to
four minutes of time. Accordingly, supposing one place be situated
10° east of another, the difference of time between them will be 40
minutes; and if it be twelve o'clock at the place which is eastward
of the other, it will be only 20 minutes past eleven in the morning at
the place situated more westerly.

PROBLEM XIII.

To find, by a Terrestrial Globe, the Bearings of Places to each other in
respect to the Cardinal Points of the Compass.

Suppose the two places be Europe and Persia, which appear
conspicuously on the globe, fig. 1, pl. 32; Europe being nearer to
the north-pole, and more towards the western side of the earth, is
found to be north-west of Persia. Or bring one of the places to the
zenith point of the globe, and place the quadrant of altitude over it;
which being brought down to the horizon, will show the apparent
bearing. The real bearing of places is that a vessel would steer, and
differs from this, on account of the variation of the compass at different parts of the earth.

PROBLEM XIV.

To find the Sun's Declination for a given Day at a given Place.

Our globe revolving round the sun on an inclined axis, a change
of situation arises, in respect to that luminary, to all parts of the
earth at different seasons. To ascertain the sun's situation at a particular place on a given day, we look on the horizon of the globe, or
in an ephemeris for his place on that day, and find it on the globe.
Then bringing the sun's place to the brass meridian, we count the
degrees between that and the equator; which shows the sun's de-

clination for that day at that place. The greatest declination of the sun north and south of the equator is about 23° 30′

PROBLEM XV.

To find the Declination of the Sun on the Celestial Planisphere, Figs. 1, 2, Pl. 29.

You perceive by the names of the months on the left side of the exterior circle, or ecliptic, fig. 1, that the sun is on the north side of the equator during the whole months of April, May, June, July, August, and part of March and September; and by the months on the right side of the ecliptic of fig. 2, that he is on the south side of it during October, November, December, January, February, and part of September and March. The sun's declination at any time may be known by fig. 1 and 2, by measuring the distance between his place in the exterior circles, and the equator A B for the given day, on the opposite extremity of the parallel of longitude on which he is then situated; for as much as the ecliptic is there distant on one side of the equator, such will be the sun's distance on the other side of it, at that time; which may be ascertained by the scale of degrees at the sun's greatest declination.

PROBLEM XVI.

To find the Places the Sun is Vertical to on a given Day.

The sun's apparent path being confined to the torrid zone, which is situated between the tropics, he is never vertical to any spot without those limits. But to the inhabitants of this zone he is vertical twice a year, and to those of the tropics once in that period.

Having found the sun's declination, bring that place to the meridian; and all the places which pass under the same degree of the meridian with the sun's declination, have the sun in their zenith when it is twelve o'clock with each of them. The cause of this, the next problem will explain.

PROBLEM XVII.

To Rectify the Terrestrial Globe to the Sun's Declination.

Elevate the pole as many degrees as the sun is distant from the equator. If his declination be north, elevate the north pole; if south, the south pole. This will bring the sun's place to the zenith point of the globe, and show that he is in the zenith, or vertical to every place so situated as to pass under that vertical spot.

PROBLEM XVIII.

To exhibit all the Places which have the Light of the Sun during a given Day with London, and those which have not.

Rectify the terrestrial globe for the sun's declination, as directed in the last problem, and turn London to the western edge of the globe, and then to the eastern edge of it, which will be the time of day at London; when you will perceive all the places on the earth that have daylight with London on that day, and those that have not.

PROBLEM XIX.

To find all the Places that see the Sun when it is twelve o'Clock at Noon with London on a given Day.

Rectify the terrestrial globe for the sun's declination for the given day, and bring London to the meridian: for bringing that or any other place to the meridian of a terrestrial globe, shows twelve o'clock at the place brought to the meridian. Suppose London to be at the meridian, then all the places above the horizon are those on which the sun shines at that time. The earth being a spherical body, only one-half of it can be illuminated by the sun at the same time.

PROBLEM XX.

To find, by Inference, the Places the Sun is Vertical to on any Day.

Knowing, by an ephemeris or globe, the sun's declination for a given day whatever it may be, the places having that latitude will have the sun in their zenith when it is twelve o'clock with them.

PROBLEM XXI.

To find the nearest Distance of Places from each other in Miles.

A degree of the equator contains 60 geographical miles, or $69\frac{1}{4}$ English. As the measure of distances varies in different countries, it is better to estimate by the geometrical quantities contained in a degree at the equator. It is evident on inspection of either the terrestrial globe or planisphere, that as the circles parallel to the equator become less and less in approaching the poles—all circles, large or small, being divided into the same number of parts or degrees—a degree of the smaller circles must contain a less quantity than one at the equator. But this does not prevent our calculating the distances of places on either of these circles; even without reference to the tables of quantities and latitudes: because, by taking their distances by a pair of compasses, and applying the same to the equator, whatever number of degrees of it is included within the distance taken, when multiplied by 60, will give the real distance.

PROBLEM XXII.

To find those Places which have Noon on a given Day when it is any other Hour at London.

Rectify the terrestrial globe to the sun's declination. Bring

London to the meridian, and turn it to the given hour. Then all the places that are under the meridian at that hour will have the sun on their meridian when it is the given hour at London.

PROBLEM XXIII.

*To find what two Days in the Year the Sun will be Vertical to a par-
ticular Place within the Torrid Zone.*

Bring the place to the meridian, and observe which two points of the ecliptic circle pass under that degree of the meridian which is equal to the latitude of the place; then refer to the horizon, to see what two days in the year the sun will be in those two points of his apparent revolution, which will be the days in which the given place will have the sun in its zenith.

PROBLEM XXIV.

*To find, by Inference, the two Days in the Year on which the Sun will
be Vertical to a given Place within the Torrid Zone.*

Knowing the latitude of the place, find by an ephemeris or a globe the two days when the sun's declination will be exactly the same as the latitude of that place.

PROBLEM XXV.

*To observe the Circumstance arising from two particular Situations of the
Inhabitants of our Earth, in which they are called Antipodes.*

The antipodes are situated exactly in opposition to each other; as thus: suppose one situated at 20° northern, and the other at 20° southern, latitude; one at the first degree of longitude, and the other at 180°; as represented by P P, fig. 1, pl. 32. The consequences arising from these relative situations are, that with one it will be twelve o'clock at night, while with the other it is twelve o'clock at noon, because they are on opposite or contrary meridians; one will have summer while the other has winter, because they are situated on different sides of the equator; and consequently the days will be longest with one when shortest with the other.

PROBLEM XXVI.

To illustrate the Circumstances of the Perioeci; or those which are situated on the same Parallel of Latitude, but differ from each other 180° in Longitude.

Suppose two places so circumstanced to be, one at the first degree of longitude, and the other at 180°, and both on the same side of the equator, in 40° northern latitude, as at R R, in fig. 1, pl. 32. These will have their seasons the same, because they are both on the same side of the equator: their times of day and night will be exactly contrary to each other, as they are on opposite meridians; but each will have the same length of day and night, because they have the same latitude.

PROBLEM XXVII.

To illustrate the Circumstances of the Antioeci; or those situated on opposite Parallels of Latitude, but having the same Longitude.

The antioeci being on opposite sides of the equator—say each 20° of latitude, as at * *—differ in their seasons, and in the length of their days and nights, at the same time; one having winter while the other has summer, and the day being longest with one when shortest with the other. They have twelve o'clock at noon and twelve o'clock at night together, because they are situated under the same meridian.

PROBLEM XXVIII.

To find what Parallel the Sun is beginning to Shine at constantly, on a given Day, within the northern Polar Circle; and from what

Parallel he is going to absent himself within the southern Polar Circle.

Our earth, revolving on an inclined axis, has sometimes one and sometimes the other polar region turned towards the sun; and the days in these parts of our globe continue for months, and the nights the same. Having found the sun's declination for the given day*, elevate the north pole the same number of degrees; and the parallel that touches the north part of the horizon will be that to which he is going to shine constantly, while the parallel at the southern part of the horizon and just below it will show the one he is going to leave in darkness.

PROBLEM XXIX.

To ascertain the same Circumstances of the southern Polar Regions.

Fix on some day between September and March when the sun is in his southern range, and elevating the south pole, observe the parallel just above the southern edge of the horizon, for there the sun is beginning to shine constantly; and from the parallel just below the northern edge of the horizon the sun is withdrawing his diurnal influence.

PROBLEM XXX.

To find the above-stated Circumstances by Calculation.

Knowing the sun's declination, either north or south, for the given day, subtract that quantity from 90°, and the remainder will be the latitude of the places to which these circumstances, depending on the sun's declination and situation in respect to the equator, belong.

PROBLEM XXXI.

To find the Time that the Sun will continue to shine, or remain above

---

* The day fixed upon must be between March and September.

T T

the Horizon, at a given Parallel of Latitude within the Polar Regions.

Suppose we say latitude 80° north, which will be when the sun has 10° of northern declination. Remarking the two points of the ecliptic that pass under 10° of northern declination, and referring to the horizon of the globe for the number of days and minutes between these two fixed periods northerly, we find the length of the longest day at that parallel; which will be the same as the length of the longest night at the correspondent one at the southern polar regions.

PROBLEM XXXII.

To find the Length of the longest Day at a given Place in the southern Polar Regions.

Proceed agreeably to the foregoing illustration; only elevating the southern pole, and fixing on some time when the sun is on the southern side of the equator.

PROBLEM XXXIII.

To find, by Calculation, what Number of Miles is contained in a Degree at a given Parallel of Latitude; the Proportion a Degree at that Parallel bears to a Degree of the Equator being known.

The proportion that a certain number of degrees at the equator bears to the same number of degrees at a certain latitude, such will be the proportion that the number of miles bear to each other.

PROBLEM XXXIV.

To observe what Places have Noon at the same Time with each other on a given Day. Suppose, for Example, all those that have Noon at the same Time with Calcutta.

Bring Calcutta to the meridian, and rectify the globe to the sun's declination; then observe all the places under the meridian, above the horizon, from the north to the south point of the compass; for these will have noon at the same time as Calcutta, and with each other, on that day.

PROBLEM XXXV.

To know what Places within the torrid and temperate Zones have twelve o'Clock at Night when it is twelve o'Clock at Noon in London, on a given Day.

Rectify the terrestrial globe as in the last problem; and after bringing London to the meridian, fix the index to twelve o'clock; then turn the globe on twelve hours, and the places that are under the meridian will be those that will have twelve o'clock at night when the sun is on the meridian of London on the given day.

PROBLEM XXXVI.

To find the Places that have Noon at the same Time with London, and those that have twelve o'Clock at Night when it is Noon at London, on a given Day, by the Planisphere.

Observe the meridian of London, and the places immediately at it, in the two projections figs. 1 and 2, pl. 35, allowing for the sun's declination on the given day; and you will see those which will have noon at the same time with London. Those which will have twelve o'clock at night at the same time are situated on the opposite meridian to London, and may be ascertained in the same way.

PROBLEM XXXVII.

To find, by a terrestrial Globe, all the Places to which the Sun is rising or setting, or it is Day or Night, at a given Hour and Day with London.

When any place is brought to the brass meridian of the terrestrial globe, that represents twelve at noon at that place, with astronomers called the first hour of the day. Rectify the globe to the sun's declination for the given day; then bring the given place, as London, to the meridian, and observe all the places lying under the meridian from the north to the south points of the horizon, for these will have the sun on their meridian at the same time with London; while with those under the opposite one it will be twelve o'clock at night, within the torrid and temperate zones. The whole part above the horizon will have day at that time; and all below it, excepting within 18° of the horizon, will have night. To the places within the torrid and temperate zones, on the western edge of the horizon, the sun will be rising; to those within the same distance on the eastern edge of it, he will be setting.

To find the Elevation of the Sun at any Place on the terrestrial Globe, at a given Hour, on a given Day.

Rectify the globe to the sun's declination for that day and hour, and observe by the quadrant of altitude what height the place at which the sun's elevation is required is above the horizon, for that will be the height of the sun above the horizon of that place at that hour on that day.

PROBLEM XXXIX.

To find the Angle of Position between two Places.

Bring one of the places to the zenith point of the globe; then bring the quadrant of altitude over the other place, and down to the horizon, and you will see the angle, or whole difference, in their position, because the one in the zenith will intersect the east or western points of the horizon; and by observing where the other intersects it, you will see the angle formed between the places on the horizon, which is called the angle of position.

An Explanation of the Relation between the terrestrial and celestial Globes.

When we elevate the pole of the terrestrial globe to the latitude of a place, that place is brought to the zenith point, as at z, fig. 1, pl. 32. When we elevate the pole of a celestial globe to the same latitudes, the circles of the heavens delineated on the celestial globe having the same position as they have on the terrestrial globe, and their poles being in the same situation with each other, an inhabitant of the earth is supposed situated as at p, on the globe t, within the armillary sphere fig. 2, pl. 33. This sphere being rectified to the latitude of London, that place is supposed to be at p, the zenith point of the globe which is to represent the earth; and the circles of the sphere, denominated the equator, ecliptic, tropic, polar and meridianal ones correspond with those of the earth; and the poles, n o, of the sphere are a continuation of those of the heavens. From this explanation, it is easy to understand that the celestial globe may be so rectified as to exhibit the circumstances of celestial phenomena as seen by the inhabitants of the earth.

PROBLEM XL.

To rectify the celestial Globe to the Sun's Place, on a given Day, for a given Latitude.

Look for the sun's place for the day on the horizon of the celestial globe, then find it on the ecliptic circle, bring it to the meridian, and fix the index to twelve o'clock. Look for the latitude of the place on the terrestrial globe, and elevate the pole of the celestial one to that latitude. Then will the celestial globe represent the position of all the circles and stars in the heavens in respect to a spectator at twelve o'clock at noon on that day, situated in the same line of direction with the zenith point of the celestial globe, and on the zenith point of the terrestrial one.

PROBLEM XLI.

To observe the Situation of a right Sphere, and the Circumstances attending those who are thus situated on our Earth.

When the poles are in the horizon, the globe represents the situation of the inhabitants of the equator: for, as they have no latitude, the poles have no elevation with them; consequently the equator and the sun's apparent diurnal motion are in the same direction with each other, as from A to B, in fig. 2, pl. 30. The inhabitants of the equator must always see the sun perform a complete semicircle above the horizon; hence the days and nights are each twelve hours long all the year round to the inhabitants of the equator. To illustrate this position by the globe, we place the poles in the horizon; and on fixing the index to twelve o'clock, and marking the sun's place for any day in the year, we shall perceive that spot will be exactly twelve hours above and twelve hours below the horizon of the inhabitants of the equator, because they are situated in a right sphere; and that his apparent revolution will be performed in a semicircle, rising and falling at right angles to the horizon.

PROBLEM XLII.

To illustrate the Circumstance of a parallel Sphere.

This position of the earth represents the inhabitants of the poles. Their latitude being 90°, the pole is elevated 90°, and the sun appears to perform entire revolutions above their horizon; his apparent diurnal motion being parallel to A B, fig. 3, pl. 30: so that the inhabitants of the poles have six months day and six months night, as may be understood from inspection of fig. 3, pl. 30; in which one half of the sun's apparent path is supposed to be above A B, and the other half below it. Thus, while one pole has six months day, the other has six months night: the upper half of fig. 3 representing day, the lower half night. This may be seen also by the globe or armillary sphere, by placing either pole in the zenith point of the horizon.

PROBLEM XLIII.

To illustrate the Circumstances of the Inhabitants of an oblique Sphere.

Fig. 4, pl. 30, represents this sphere, with the north pole elevated to the latitude of London. It is evident, from the situation of the ecliptic, that the inhabitants of an oblique sphere see the sun when he is on each side of the equator; but that he makes the largest angle to the inhabitants of the north temperate zone when he is on the northern side of the equator, and the least when on the southern side of it. Hence it is that the days are longest with us when the sun is in the tropic of Cancer, and shortest when he is in the tropic of Capricorn: and *vice versa* to the inhabitants of the southern temperate zone; for they have their shortest day when ours is at the longest, and their longest at the time of our shortest.

Let A, fig. 4, represent the situation of a spectator at London: it is evident that the sun is less oblique to that spot when at B, the tropic of Cancer, than at C, the tropic of Capricorn. Hence his rays fall more abundantly on us in the former than in the latter situation: and, on account of the greater angle he makes when at Cancer, the days are then longest with us, and we have summer; while the southern temperate regions have winter: and *vice versa*. This may be illustrated by placing the globe, or armillary sphere, in the horizon to our latitude, and remarking the time that the sun is above the horizon at each situation. Bring the sun's place, when in the tropic of Cancer, to the meridian, and fix the index to twelve o'clock, as before directed. Then turn the globe to the eastern, and afterwards to the western, edge of the horizon of a celestial globe, to observe the times of the sun's rising and setting by the index. Rectify the celestial globe to the sun's place for the tropic of Capricorn; pursue the same mode, and make the same observations, and also for other days in the year, and you will find that the day is longest to us when the sun is in the tropic of Cancer, and shortest when he is in the tropic of Capricorn.

PROBLEM XLIV.

To find by a terrestrial Globe the Part of the Earth where an Eclipse of the Sun will be seen, within a certain Distance of the zenith Point*.

Finding in the ephemeris the day and hour of a solar eclipse, also the sun's place on that day, rectify the terrestrial globe to the sun's declination, and bring London to the meridian. Observe the degree of the meridian at the sun's declination, and turn the globe to the given hour of the eclipse; when the inhabitants of places under the meridian at that time, at the place of the sun's declination, will see the sun eclipsed in their zenith; and, if the eclipse be total, more or less of it will be visible to within about 80° from that spot; if partial, not to so great a distance. Thus we are not able constantly to ascertain by a globe all the places where an eclipse of the sun will be seen; for sometimes it may be visible to only a very small portion of the upper hemisphere of the globe. As the moon's shadow hides but a small portion of the sun in respect to the earth, an eclipse of the sun may not be seen at the greater number of the places above the horizon that have his light at the time; for which reason also we cannot, by a celestial globe, authenticate the predicted circumstances of an eclipse of the sun.

* For a general idea of the cause of eclipses, see page 279, Lecture the thirteenth of this book; and for a fuller account, see my Astronomy.

PROBLEM XLV.

To find all the Places on the Earth to which a predicted Eclipse of the Moon will be visible.

Rectify the terrestrial globe to the sun's declination and the meridian of London; turn it on to the given hour, and see what spot is immediately at the place of the sun's declination. Then turn the globe to the opposite part to that, which will represent the moon's situation at that time; when the eclipse will be visible at all places above the horizon of the globe.

PROBLEM XLVI.

To find on a celestial Globe at what Time an Eclipse of the Moon will be first seen by us.

Rectify the celestial globe to the latitude, meridian and sun's place. Look for the moon's place in the ephemeris, and find it on the globe; fix a patch on it, and, turning the globe naturally, observe the predicted hour of our seeing it, and notice if it be above the horizon and the sun below it.

PROBLEM XLVII.

To find the Sun's meridional Altitude for a given Day and Place.

On account of the obliquity of the ecliptic to the equator, the sun's altitude above the horizon of a place varies each day. By the observation of this circumstance, the declination of the sun was ascertained, and laid down on globes and maps. Rectify the celestial globe; then count the degrees between the horizon and the sun's place on the meridian, which will be the sun's meridional altitude, or greatest height above the horizon, for that day at that place.

PROBLEM XLVIII.

To find the Sun's Altitude for any Hour of a given Day, at a given Place ; also his Bearing in respect to the Points of the Compass.

Rectify the celestial globe; then turn the sun's place to the given hour, apply the quadrant of altitude to the zenith point of the globe, and bring it over the sun's place and down to the horizon. The number of degrees between the sun's place and the horizon will be the height of the sun at that time, at the given place; and the point of the compass on the horizon intersected by the quadrant will be its bearing.

PROBLEM XLIX.

To find the Altitude and Bearing of a fixed Star above the Horizon at a given Time, on a certain Day, at a given Latitude.

For a place in northern latitude, fix on any star not within that distance from the south pole that is equal to the latitude of the place of observation. Then rectifying the celestial globe, turn it till the index points to the given hour; which will show the place of the star at that hour on that day. Its altitude and bearing may be taken in the manner expressed in the last problem.

PROBLEM L.

To find the Latitude and Longitude of the Stars.

The longitude of the heavenly bodies is counted on the ecliptic, and their latitude from the ecliptic, of a celestial globe. Hence it is that a celestial globe has poles to the ecliptic, on which to fix the quadrant of altitude, in order to calculate the distance of a fixed star from that circle. The degree of the ecliptic intersected by the quadrant at the same time is its longitude, counted from the equinoctial point Aries to the east and west.

PROBLEM LI.

To find the Latitude and Longitude of a Planet on the Globe.

The latitude and longitude of the stars are always the same, or nearly so; but the latitude of the planets, on account of their annual motion, is continually varying.

Find the place of the given planet in the ephemeris, within the eight parallels drawn on each side of the ecliptic circle, which will ascertain its latitude and longitude.

PROBLEM LII.

To find whether a particular Planet will be above the Horizon after Sun-set on a given Night, at a given Place, also its Altitude and Bearing at any Hour of the Night.

Rectify the celestial globe; and affix a patch to the planet's place on it. Turn the globe from east to west till the sun sets; and observe if the planet be above the horizon; for if it be, you will see it. Then applying a quadrant of altitude to the zenith point of the globe, and bringing it over the planet's place, and down to the horizon, you may ascertain its altitude and bearing at that time, or for any other when it will be above the horizon, while the sun is below it during the different hours of the night.

PROBLEM LIII.

To find which Hours of the Twenty-four the Moon will be above the Horizon at a given Latitude, on a given Day.

Find the moon's place, also its latitude and longitude in the ephemeris. Then rectify the celestial globe, and observe by the index what hour the moon will appear above the eastern edge of the horizon, and also what hour she will set to the western edge of it.

PROBLEM LIV.

To find the Declination of a fixed Star, or its Distance from the Equator.

The fixed stars never alter their declination.   Hence, if we bring one of them to the meridian, and observe its distance from the equator, we find its constant declination.

PROBLEM LV.

To find the Declination of the Sun, Moon, or a Planet, for a given Day.

Find the planet's place in the ephemeris, and then put a patch on the globe to represent it in that place.   Fix the quadrant of altitude to the zenith point of the globe, bring it to the equator, and observe the declination of the given planet for that day, or its distance from the equator.

PROBLEM LVI.

To find the Sun's Almacanther, or Circle of Illumination, above the Horizon, for any Hour.

This is merely observing the sun's height above the horizon; the circle drawn parallel to the horizon at that height being its almacanther at the time. This is the same as the parallel of altitude, and is more generally so expressed by modern astronomers.

PROBLEM LVII.

To find the Sun's Amplitude, or Distance from the East Point of the Horizon, at rising, and from the West at setting.

As the sun's apparent path is sometimes more northerly, and at others more southerly, he rises and sets nearer to the north or south points of the horizon according to these circumstances.

Rectifying the celestial globe, observe the degree of the horizon on the eastern side of it at which the sun rises; and the distance that degree is from the east point is called his amplitude. The sun's distance from the west point in setting is also called the same, and is equal to that at rising. The sun's greatest amplitude, at the latitude of London, north and south, is 40°.

PROBLEM LVIII.

To find the two Days of the Year when the Sun rises due East and sets due West to the Inhabitants of the Earth situated in the Torrid and Temperate Zones.

Having found the equinoctial points, or those degrees of the ecliptic in which the sun intersects the equator, then by looking at the correspondent days and degrees on the horizon of the globe you will find the days required.

PROBLEM LIX.

To prove that the Sun rises due East and due West only on the two Days in which he is in the Equator.

Rectify either the globe or the armillary sphere to the sun's declination for any other day within the latitude of $66\frac{1}{2}°$, and you will find that the sun's place will not intersect the horizon at the east and west points of it.

PROBLEM LX.

To prove that the Inhabitants of the Equator have always twelve Hours Day and twelve Hours Night all the Year round.

Fix on any day of the year, and place the poles of the globe or armillary sphere in the horizon; and you will find by revolving the globe or sphere, and observing the dial, that the part of the ecliptic which represents the sun's place for any day, will be exactly twelve hours above and twelve below the horizon of the inhabitants of the equator.

To find the Length of the longest Day and shortest Night at any Place within the Torrid and Temperate Zones, not situated directly on the Equator.

If the place be in northern. latitude, rectify for the sun's place when in the tropic of Cancer; if in south latitude, when he is in the tropic of Capricorn. After rectifying the celestial globe, turn the sun's place to the eastern edge of the horizon for the time of his rising, and to the western edge for the time of his setting, and you will find the length of the longest day for that latitude. Then subtracting the length of that day from twenty-four hours, the length of the night at the same latitude will be determined.

To find the Length of the shortest Day at a Place within $66\frac{1}{4}°$ North of the Equator.

Rectify the celestial globe for a place in northern latitude as in the last problem, for the sun's place in the tropic of Capricorn, and proceed as in that example.

To find, by Inference drawn from the last two Problems, the Length of the shortest Day and longest Night at any given Place.

Whatever be the length of the shortest night is the length of the shortest day at a given place; and the length of its longest day the length of its longest night.

To find by the Length of the longest Day at a given Place what Climate it is in between the Equator and the Polar Circles.

The portions of the earth between the equator and the polar circles are each divided into twenty-four parts, or divisions of half hours, called climates, and at the equator the days are always twelve hours long. Hence if we have the length of the longest day at any other parallel, we may ascertain what climate it is in. As London, for instance, where the longest day is sixteen hours and a half, the difference of time between the length of the day at the equator and the longest day at London being four hours and a half, London must be situated in the ninth climate.

PROBLEM LXV.

The Length of the longest Day at a Place in northern Latitude being given within $66\frac{1}{2}°$ of the Equator, to find its Latitude.

Without rectifying the globe to any latitude, bring the tropic of Cancer to the meridian; and fixing the index to twelve o'clock, turn the globe on half the number of hours for the given day. Then move the meridian till the tropic of Cancer touches the western edge of the horizon, and observe the elevation of the north pole, for that must be the latitude of the place.

PROBLEM LXVI.

Knowing the Length of the longest Day at a given Parallel of Latitude, to find by Calculation in what Climate that parallel is situated.

Suppose the length of the day be twenty hours, subtract twelve for the length of the day at the equator; eight hours being sixteen half hours, we find the place is situated in the sixteenth climate from the equator.

PROBLEM LXVII.

To know at what Points of the Compass the Sun will rise and set on a given Day at a given Latitude.

Rectify the celestial globe, and observe what point of the compass at the eastern edge of the horizon the sun rises at, and the point at the western edge to which he sets.

PROBLEM LXVIII.

To know what Hours of a given Day, at a given Latitude, the Sun will be due East and due West.

Rectify the celestial globe. Fix the quadrant of altitude on its zenith point, and bring it down to the eastern edge of the horizon. Turn the sun's place back eastward, and observe by the index the hours it touches the quadrant. Then fix the quadrant, so as to bring it to the western edge of the horizon; and turn the sun's place till it touches the quadrant, making the like observation as before.

PROBLEM LXIX.

To find the Day of the Month at a given Latitude, by the Observation of the Sun's meridional Altitude.

Elevate the pole of the globe to the latitude of the place, then bring the part of the ecliptic which has exactly the height above the horizon to the meridian, which agrees with the observation previously taken by a quadrant; and by looking for its correspondent degree on the horizon of the globe, the day of the month will be found.

PROBLEM LXX.

To find the Sun's Altitude for any Hour of the Day at a given Latitude.

Rectify the celestial globe, and turn the sun's place to the given hour, as shown by the index. Then apply the quadrant of altitude to the zenith point of the globe, bringing it over the sun's place down to the horizon, and count the height of the sun's place above that circle.

PROBLEM LXXI.

Having taken, by a Quadrant, the Sun's Altitude on a given Day, at a given Latitude, to find the Hour of the Day.

Rectify the celestial globe, and turn the sun's place till it have the altitude previously observed : then look at the index for the hour.

PROBLEM LXXII.

To find the Duration of Twilight at a given Place on a given Day.

The approach and departure of the sun are happily moderated in their effects to us by the refracting power of the atmosphere encompassing the earth; this effect produces the crepusculum, or twilight, which begins in the morning when the sun is about 18° below the horizon, and continues of an evening till he is sunk the same number of degrees below it.

The duration of twilight varies at different seasons at the same place, according as the sun's apparent path through the 18° below the horizon is more or less oblique to that circle. It also differs at different latitudes. In the table of diurnal periods in the ephemeris, the length of the days is reckoned from the beginning of the morning twilight, or Aurora, to the end of the evening twilight. We find, by this table, that at London from about the 24th of May to the 24th of July it is all day-light, because the sun does not sink more than 18° below the horizon of London during that time. To find the duration of twilight at London at any other time, we rectify the celestial globe for any day between the 24th of July and the 24th of May, and fix the quadrant on the zenith point of the globe; when the degrees of the quadrant below the horizon will show when the sun's place is within 18° of the horizon, both before rising and after setting.

X X

and by looking at the index, we perceive the time of the sun arriving at that spot. First turning the globe to the eastern edge of the horizon, and remarking the time that the sun's place arrives at 18° below it, and also when he transits the eastern edge of the horizon, we find the duration of morning twilight; and as the evening twilight will be the same, we can thus estimate the whole duration of twilight on that day at London.

PROBLEM LXXIII.

To find the Duration of Twilight at either Pole, after the Sun has disappeared below the Horizon of that Pole.

Though it is said that at the polar regions there is continual day for six months, and continual night for the same time; yet, if we allow for the duration of twilight, we shall find there is not quite three months total darkness at the poles.

Rectify the celestial globe to the latitude of either pole, and observe that point of the ecliptic in the opposite hemisphere that is within 18° of the horizon; and we shall find that there will be nearly two months twilight at the given pole, while the sun is increasing his declination from the equator, and nearly two months when he is approaching that circle again. We shall also find total darkness to the north pole begins about the 12th of November, when the sun is in 18° of southern declination; and that it ends at the north pole about the 30th of January: at the south pole, total darkness begins when the sun is in 18° of northern declination, and ends when he again arrives within the same distance of the equator. The degree of the meridian below the horizon, and the sun's places, will show these circumstances, by bringing the sun's places to it. The moon shines on the poles even when the sun's influence ceases; and the Aurora

Borealis also affords considerable illumination at the north pole. The foregoing problem may be satisfactorily performed by the armillary sphere.

PROBLEM LXXIV.
To ascertain the Stars which never rise, also those which never set, at a given Latitude.

Rectify the celestial globe to the given latitude, and then turning it round on its axis, these stars will be seen.

PROBLEM LXXV.
To know the Portion of the Celestial Sphere always above the Horizon of a Place, and that always below it, by Inference.

Those living within that distance of the pole which is equal to the latitude of the place in the same hemisphere, always have the stars situated within that distance of that pole above their horizon, and never see those that are within the same distance of the contrary pole.

PROBLEM LXXVI.
To find if a particular Star of those which do rise and set to the Horizon of a given Place will be above it on a particular Night.

Rectify the celestial globe, and turn the sun on to his setting; and observe if that star be above the horizon during the night, by bringing the sun's place round to his rising the next morning.

PROBLEM LXXVII.
To find the Right Ascension of the Sun, or any heavenly Body, at London.

Right ascension means that degree of the equator which comes to the meridian of an oblique sphere with the sun or star, counted from Aries. In a right sphere, it is the same degree of the equator that rises with the sun or star; and hence it is called the right ascension, even in reference to an oblique sphere also.

Rectify the celestial globe to the latitude of London, and observe what degree of the equator, counted from Aries, is at the meridian with the sun's place for the given day, for that is its right ascension.

PROBLEM LXXVIII.
To find the Right Ascension of the Sun to the Inhabitants of the Equator.

Rectify the celestial globe for the inhabitants of the equator. Then bringing the sun's place to the eastern edge of the horizon, the degrees of the equator at the sun's place will be its right ascension.

PROBLEM LXXIX.
To find the Oblique Ascension and Descension of the Sun, or a Star, on a given Day, at a given Place.

Rectify the celestial globe for any place in an oblique sphere, and observe the degrees of the equator at the eastern edge with it, which will be its oblique ascension. For its oblique descension, turn it on to the western edge of the horizon, and remark the degree of the equator with it at setting, for that will be its oblique descension.

PROBLEM LXXX.
To find the Ascensional Difference.

This is the difference between the oblique and right ascension to any place in an oblique sphere, which, on being converted into time, shows how much the sun rises and sets before or after six at a given latitude. The sun always rises to the inhabitants of the torrid and temperate zones before, and sets after six, when he is on the same side of the equator on which they are situated; and rises after, and sets before six, when he is on the contrary side of it from them.

PROBLEM LXXXI.
To find the Sun's Azimuth at Rising.

Rectify the celestial globe, and observe how many degrees there are between the north or south point of the horizon at the sun's rising in the east, for that is called his azimuth.

PROBLEM LXXXII.
To find the Sun's Azimuth and Bearing, in respect to the Points of the Compass, at any Hour of the Day.

Rectify the celestial globe, and fix the quadrant to the zenith point; then turn the globe to the given hour, and bringing the quadrant over it, observe at the horizon its distance from the north or south points, and also the points of the compass the quadrant intersects.

PROBLEM LXXXIII.
To find the Places of the Moon's Nodes.

Look in the ephemeris for the moon's ascending node, and find that place on the globe. Then look for the opposite point to that, which will be her descending node. The moon's ascending node is set down in the ephemeris for every sixth day *.

---

* Only the ascending node, or moon's departure from the ecliptic northerly, is set down in the ephemeris; because the other may be thus readily found, being directly opposite to it. The motion of the moon's nodes, through all the signs of the ecliptic, though continual, is very slow, for the moon is nearly nineteen years in passing through them.

The place of the moon's nodes, or the place where her path crosses the ecliptic, is continually changing. The cause of this is explained in my Lectures on Astronomy.

PROBLEM LXXXIV.

To know the Situation in the Heavens of the Moon's Orbit at any Time.

Pass a ribbon round the globe, confining it in two opposite points, called the nodes of the moon's orbit. The moon's greatest distance from the ecliptic is not more than 5° 15′ on each side of it. Her orbit may be seen better by tying a string round the armillary sphere than round the globe.

PROBLEM LXXXV.

To find the Place of the Moon's Latitude and Longitude on the Globe;
also the Time of her rising and setting in respect to a given Lati-
tude on a given Day.

Look for the latitude and longitude of the moon in the ephemeris. Find these on the globe; that is, her place in respect to the ecliptic, and distance from it. Then rectify the celestial globe, and bring the moon's place to the eastern edge of the horizon, and observe the time of her rising; the same when she is at her meridian, for her southing; and the time when she is at the western edge of the horizon, for the time of her setting.

PROBLEM LXXXVI.

To find the Declination of a Planet on a given Day.

Find the place of the planet among the eight parallels to the ecliptic; then bring it to the meridian; and the degrees between the equator and the place of the planet will be its declination.

PROBLEM LXXXVII.

To find the Moon's greatest Meridional Altitude at London.

Bring 5° 15′ of the ecliptic, which is the moon's greatest latitude and distance from the tropic of Cancer, to the meridian, and fix the globe to the latitude of London. The degrees between that spot and the horizon will show the greatest altitude of the moon at that latitude, which is 5° 15′ more than the greatest altitude of the sun at that place.

PROBLEM LXXXVIII.

To find the Sun's greatest Altitude at London.

Rectify the globe to that latitude; and when the tropic of Cancer is at the meridian, the difference between that and the south pole of the horizon, counted on the meridian, will be about $62\frac{1}{2}$ degrees for the sun's greatest meridional altitude at London, whereas that of the moon is about $67\frac{1}{4}$ degrees.

PROBLEM LXXXIX.

To find the Moon's least Meridional Altitude at London, and its Situation as to the Ecliptic at that Time.

Rectify the celestial globe, and bring the 28° 47′ on the south side of the equator, at the tropic of Capricorn, to the meridian, which is the moon's greatest declination; and count the degrees from that to the southern point of the horizon. This will show the moon's least altitude at London.

PROBLEM XC.

To find the Sun's least Meridional Altitude at London.

Rectify the celestial globe, and bring Capricorn to the meridian; and count the degrees thence to the southern part of the horizon The moon's least altitude is less than that of the sun.

PROBLEM XCI.

To find the Hours of a Star Rising, Culminating and Setting, at a given Latitude on a given Day.

Rectify the celestial globe, and turn it till you see the star rising, and look at the index. Turn it on, in like manner, to the meridian, and observe the time of its culminating: then on to the western edge of the horizon, for the time of its setting at that latitude.

PROBLEM XCII.

To find all the Stars that are Rising, Setting, or on the Meridian, at a certain Hour of the Night; and all those above the Horizon of a given Place.

Rectify the celestial globe; turn it on to the given hour; then all the stars at the eastern edge of the horizon will be rising, those under the meridian of the globe will be on the meridian of the given place, and all on the western edge of the horizon will be setting to it; and the whole celestial hemisphere above the horizon may be seen at that time at that place.

PROBLEM XCIII.

To find the Day in the Year that will be of the same Length with a given Day at a given Latitude.

Observe the degree of the meridian intersected by the sun's place in the ecliptic, which will be its declination for the given day. Turn the globe on till another degree of the ecliptic passes under the same degree of the meridian; when, referring to the horizon, you will find the day corresponding with the sun's place, which will show the day in the year that will be exactly the same length with the given day.

PROBLEM XCIV.

To find by the Altitude of a Star at a given Latitude the Hour of the Night.

Take the height and bearing of the observed star by a quadrant and compass; then rectify the celestial globe, and turn it on till the star has the same elevation and bearing on the globe as it has in the heavens; then looking at the index, you will find the hour of the night.

PROBLEM XCV.

To find the Hour of the Day by the Sun's Azimuth.

Observe the sun's azimuth, or bearing towards the north or south point of the compass. Rectify the celestial globe, and fix the quadrant of altitude to the zenith point; and bring it down to intersect the horizon at the point that is equal to the sun's azimuth. Then turning the globe till the sun's place intersects the quadrant of altitude, and looking at the index, you will find the hour of the day.

PROBLEM XCVI.

To find the Latitude of a Place by observing the Sun's meridional Altitude by a Quadrant.

Find the sun's place for the given day, and bring it to the meridian of the globe; elevate or depress the pole till the sun's place have the relative altitude on the meridian of the globe that it has on that circle in the heavens. Then look at the degree at the north or south point of the horizon, according as it is in north or south latitude, for the latitude required.

PROBLEM XCVII.

To find the Place of the Sun when below the Horizon for any Hour of the Night.

Rectify the celestial globe and turn the sun's place on till the index points to the given hour, when you will find the sun's place below the horizon at that time.

PROBLEM XCVIII.

To find the relative Situation of the Sun and Moon to each other at any Hour at a given Place, when the Moon is above and the Sun below your Horizon; by which you will understand why you see different Portions of the Moon's enlightened Face at different Times.

Find the moon's place in the ephemeris, and fix a patch on the globe at that situation. Next rectify the celestial globe, and turn it to the given hour; then you may observe the relative situation of the sun and moon in respect to each other, by allowing half a degree of the moon's motion each hour from twelve o'clock at noon.

PROBLEM XCIX.

To find on what Day a given Star rises with the Sun, or cosmically, at a given Latitude.

Rectify the celestial globe for the given latitude, and observe what part of the ecliptic is at the eastern edge of the horizon with the star; then look for its correspondent day on the horizon, which will show the day on which that star rises cosmically.

PROBLEM C.

To find when a Star sets cosmically, or at Sun-rise, at a given Latitude.

Turn the star to the western edge of the horizon, and observe what point of the ecliptic is rising in the east; for that will show the day that the given star sets cosmically, or at sun-rise.

PROBLEM CI.

To find when a Star rises and sets achronically, or at Sun-set, at a given Latitude.

Rectify the celestial globe for the given latitude. The degree of the ecliptic at the eastern edge of the horizon when the sun is setting in the west, shows the day of the achronical rising of that star, or when it rises at sun-set; the degree of the ecliptic that is at the western edge setting with a given star, shows the day the given star sets achronically, or at sun-set.

PROBLEM CII.

To observe when a Star rises and sets heliacally at a given Place.

The heliacal rising of a star is when, in a morning, it appears just above the eastern edge of the horizon, before the sun; and the heliacal setting is when it sets immediately after the sun is invisible at the western edge of the horizon. Rectify the celestial globe to the latitude of the given place; then observe the degree of the ecliptic that is immediately under the eastern edge of the horizon when the star is above it, for that will show the day that star rises heliacally. For its heliacal setting, turn the star to the upper edge of the horizon on the western side of it, and observe the degree of the ecliptic just below the western side of the horizon at the same time, for that will show the sun's place on the day that star sets heliacally *.

---

* The different magnitudes of the stars cause a difference in their appearance, before sun-rise and after sun-set. If the star be of the first magnitude, it may be seen at the horizon when the sun is 12° below it; if of the second magnitude, not till the sun is 13° below the horizon; the third magnitude, when he is 14°; and the fourth magnitude, not till he is 15°, below the horizon.

PROBLEM CIII.

To find, by a celestial Globe, if Jupiter or Venus be the Morning Star on a given Day at a given Place.

Find the planets' place in the ephemeris, and fix patches to show their relative situations on the celestial globe. Rectify the celestial globe; and that planet which rises on the given morning immediately before the sun, is a morning star at that time.

PROBLEM CIV.

To find, by a celestial Globe, whether Jupiter or Venus be the Evening Star at a given Latitude on a given Day.

This is performed exactly as the last problem; excepting, that the planet at the western edge of the horizon after the sun has set is the evening star.

PROBLEM CV.

To find the Hour of the Night by the Altitude and Azimuth of a Star.

Rectify the celestial globe, and turn it on till the observed star has the same altitude and bearing in respect to the horizon of the globe that it has in the heavens; then, by fixing the quadrant of altitude to the zenith point of the globe, and bringing it down to the observed azimuth, on looking at the index you will find the hour.

PROBLEM CVI.

To find by two Stars, having the same Azimuth, the Hour of the Night.

Observe two stars, and note their bearings, or azimuth. Rectify the celestial globe, and fix the quadrant of altitude to the zenith point; then bring it to the given azimuth; and when those two stars touch the edge of the quadrant, look at the index for the hour of the night.

PROBLEM CVII.

To find, by the Altitudes of two Stars, having the same Azimuth at a given Hour, the Latitude of a Place.

Take the altitudes of the two stars, and their azimuth. Then rectify the celestial globe for the sun's place for a given day; turn it on to the given hour, and move the globe north or south till the two observed stars have the same relative situations above the horizon of the globe that they have in the heavens; then, looking at the degree of the meridian at the part of the horizon under the elevated pole, you will find the latitude of the place.

PROBLEM CVIII.

To find what Night in the Year a given Star will be on our Meridian at twelve o'Clock at Night.

Bringing the star to the meridian, observe what degree of the ecliptic is at the meridian with the star; then turn the globe on till the opposite part to that in the ecliptic comes to the meridian, which will be the sun's place in the ecliptic circle on the day that star will be on the meridian at twelve o'clock at night.

PROBLEM CIX.

To find what Hour a given Star will be on our Meridian on a given Day.

Rectify the celestial globe, and turn it on till the given star comes to the meridian, and the index will show the hour.

PROBLEM CX.

To illustrate the Circumstance of the Harvest Moon.

At the autumnal equinoxes with us, or when the sun is in the first degree of Libra and the moon in Aries, the moon rises and the sun sets with very little difference of time, thus affording almost perpetual good light.

Place six patches to represent the moon on the celestial globe, beginning at Aries, and leaving 13° between each for the moon's mean daily motion in the ecliptic; and also six patches for the sun's apparent diurnal motion, which is one degree each day, beginning at Libra. Then turning the globe on for six days, allowing the time of the sun setting and the moon rising each day, the difference will be seen; and on the sixth day, there will be only two hours difference of time between the sun setting and the moon rising. This full moon is called the harvest moon, because it happens in the time of harvest with places situated north of the equator, and affords a continuance of good light during that period.

PROBLEM CXI.

To find the Longitude of a Place by the Observation of an Eclipse of one of Jupiter's Satellites.

The time of an eclipse of one of Jupiter's satellites being found in the ephemeris as calculated to be seen at Greenwich; if at any other situation this circumstance be observed at two hours later, it shows that place is 30° east of Greenwich; if it happens two hours sooner, the place is 30° west of the Royal Observatory, Greenwich.

ASTRONOMICAL AND GEOGRAPHICAL QUESTIONS AND EXERCISES: EACH REFERRING, BY A FIGURE, TO ITS CORRESPONDENT SOLUTION IN THE PRECEDING PROBLEMS.

In these Exercises are introduced the names of the principal countries, cities, oceans and seas of the known world; also the names of some of the most conspicuous stars in each constellation of the heavens, visible at the latitude of London—for the beneficial purposes of general and particular information, and to fix the recollection of the situations and circumstances of some of the most distinguished places and stars in the minds of young geographical and astronomical students.

QUESTIONS AND EXERCISES.

Figures of re-
ference to the
Problems.

On what circles of the terrestrial globe are the latitude and longitude of places counted?   1

What are the latitude and longitude of London, the capital of England?

What of Edinburgh, the capital of Scotland?

What of Dublin, the capital of Ireland?

What of Bergen, the capital of Norway?

What of Copenhagen, the capital of Denmark?

What of Stockholm, the capital of Sweden?

How do you find these on the planisphere? Plate 35.   2

How do you find what place is situated in a particular latitude and longitude?   3

What place in Russia is situated in 59° 56′ north latitude, and 30° 24′ east longitude?

What in 52° 14′ north latitude, and 21° 5′ east longitude in Poland?

Figures of re-
ference to the
Problems.

What place in Germany in 48° 12′ north latitude, and 16° 22′ east longitude?

What in Holland in 52° 22′ north latitude, and 4° 49′ east longitude?

What in Flanders in 50° 51′ north latitude, and 4° 26′ east longitude?

*How do you find all the places that have the same la-
titude?*   4

What place in France has the same latitude with Paris?

What place in Spain, with Madrid?

What in Portugal, with Lisbon?

What in Switzerland, with Berne?

*How do you find the difference of latitude between two
places situated on contrary sides of the equator?*   5

What is the difference of latitude between the Cape of Good Hope in Africa, and Constantinople in European Turkey?

What between Gibraltar at the mouth of the Mediterranean Sea, and the Bay of Bengal in the East Indies?

What between Athens in Greece, and Calcutta in the East Indies?

*How do you find all the places that have the same lon-
gitude?*   6

What places have the same longitude with the island of Iceland in the Northern Ocean?

What the same as Lapland in the Northern Ocean?

Figures of re-
ference to the
Problems.

What the same as Majorca, an island in the Mediterranean Sea?

*How do you find the difference in longitude between two
places, when both are east or west? also, when one has eastern
and the other western longitude?*   7

What is the difference of longitude between Minorca
and Corsica, both islands in the Mediterranean Sea?

What between Sardinia and Sicily, islands in the Mediterranean?

What between the Indian Ocean and the Gulph of
Mexico?

*How do you find all the places that have the same latitude
or longitude on the planisphere?*   8

What places have the same latitude or longitude with
the island of Rhodes in the Levant?

What with the island of Patmos in the same sea?

*How do you find the difference of latitude and longitude
between places on a planisphere?*   9

What is the difference of latitude and longitude between Copenhagen and London?

What between Denmark and Sweden?

*How do you find, by the terrestrial globe, the hour of the
day at different places at the same instant?*   10

What is the difference of time between Lima, the capital of Peru in South America, and Cairo in Egypt?

Figures of re-
ference to the
Problems.

What between Philadelphia in North America, and Canton in China, the Chinese port resorted to by Europeans for traffic?

What between Cape Verd in Africa, and Aleppo in Syria?

How do you find the same by the planisphere?   11

How do you find the difference of time between places by calculation only?   12

What is the difference of time between Cape Horn the southern extremity of South America, and Botany Bay in New Holland?

What between a place situated at 15° east of London, and one situated 30° west of it?

What between London and a place 45° east of it?

What between London and a place 60° west of it?

How do you find the apparent bearings of places to each other?   13

How does London bear in respect to the North Sea?

How does London bear in respect to the Baltic Sea?

How do you find the sun's declination for a given day?   14

What is the sun's declination on the 14th of March?

What on the 28th of June?

What on the 22d of December?

What on the 12th of October?

Figures of re-
ference to the
Problems.

*How do you find the sun's declination on a plani-
sphere?*

15

*How do you find the places the sun is vertical to on a
given day?*

16

Where is he vertical on the 1st of May?

Where on the 14th of August?

Where on the 7th of September?

Where the 14th of April?

*How do you rectify the terrestrial globe to the sun's
declination?*

17

How do you rectify it to the sun's declination 20° north
of the equator?

How for 20° south of the equator?

*How do you exhibit on a terrestrial globe all the places
that have the light of the sun on a given day with London,
and those which have not?*

18

Show all those which have day with London on the
20th of May.

Show all those that have day with us on the 20th of
October.

*How do you exhibit all the places on the terrestrial globe
which see the sun when it is twelve o'clock at noon at any other
place on a given day?*

19

Which places see the sun when it is twelve o'clock at
noon at Port Royal in Jamaica, on the 1st of May?

Figures of re-
ference to the
Problems.

Which, when it is twelve o'clock at **London on the 1st**
of June?

Which, when it is twelve o'clock at Bombay in the
East Indies, on the 1st of July?

*How do you find by inference the places the sun is vertical
to on a given day?*   20
When is the sun vertical to St. Christopher's Island, in
the Caribbean Sea, North America?

When to Cape Comorin in the East Indies?

*How do you find the nearest distance between places in
miles?*   21
What is the nearest distance between Amsterdam in
Holland, and London?

What between Breslaw in Europe, and the Bay of Biscay on the coast of France?

What between the island of St. John, Antigua, in the
Caribbean Seas, North America, and London?

*How do you find the places that have noon when it is any
other hour at London, or elsewhere?*   22
Which places have noon when it is ten o'clock in the
morning at London?

Which places have noon when it is one o'clock in the
afternoon at Aleppo in Syria?

Which places, when it is nine o'clock in the morning at
Alexandria in Egypt?

Figures of re-
ference to the
Problems.

Which, when it is four o'clock in the afternoon at Arch-
angel in Russia, Europe?

*How do you find on what days in the year the sun will be
vertical to a given place within the torrid zone?*   23
When will the sun be vertical to Bombay, East Indies?
When to the island of Pines in the Pacific Ocean?
When at Mecca in Arabia?
When at Palliser's Isles in the Pacific Ocean?
When at Maskelyne's Isles, Pacific Ocean?

*How do you find by inference the two days in the year on
which the sun will be vertical to a particular place in the
torrid zone?*   24
When is he vertical to the Sandwich Isles, in the Pacific
Ocean?

*How do you make observations on the different circum-
stances arising to those situated in exact opposition to each
other, called Antipodes?* Name the circumstances.   25
What spot on our earth is the Antipodes of Bagdad, in
Turkey in Asia?
What of Barcelona in Spain?
What of Cadiz in Spain?
What of Ephesus in Turkey in Asia?

How do you illustrate the circumstances of the periœci?
Name the circumstances.   26
Which are the periœci of Kingston in Jamaica?

Which of Jerusalem, Turkey in Asia?
Which of Madeira, Atlantic Ocean?
Which of Madras, East Indies?

How do you illustrate the circumstances of the antœci?   27
Name these circumstances.
     Which are the antœci of Madrid in Spain?
     Which of Malta in the Mediterranean?
     Which of Lisbon, Portugal?
     Which of Lizard Point in Cornwall?

How do you find the parallel of the north frigid zone to which the sun is beginning to shine constantly on a given day; and from what parallel in the south frigid zone he is going to absent himself?   28
     What parallel on the 12th of May?
     What on the 1st of June?
     What on the 12th of June?

How do you ascertain the parallel in the south polar regions to which the sun is beginning to shine constantly on a given day; and that from which he is beginning to absent himself constantly in the north frigid zone?   29
     What parallel on the 1st of October?
     What parallel on the 19th of October?
     What on the 1st of November?

How do you find the above-stated circumstances by calculation?   30

Figures of re-
ference to the
Problems.

On what day does the sun begin to shine constantly at the latitude of 70° north, and to absent himself at the same latitude south?

On what day at 70° south, and absent himself at 70° north?

How do you find the continuance of the longest day at a given parallel in the south frigid zone? and the length of the longest night at the same parallel in the south frigid zone?　　31

What is the length of the longest day at latitude 80° north?

What at 70° north?

What at 74° north?

How do you find the continuance of the longest day at a given parallel in the south frigid zone, and the length of the longest night at the north frigid zone?　　32

What is the length of the longest day at 69° south?

At 72° south?

At 83° south?

How do you find by calculation the number of miles contained in a degree at a given parallel of latitude?　　33

How many miles are there to a degree in latitude 30°?

In latitude 40°?

In latitude 51° 30'?

How do you find all the places that have noon on a given day with a given place?　　34

Figures of re-
ference to the
Problems.

With London on the 1st of May?

With Pisa in Italy, on the 1st May?

With the Isle of Bermudas in the Pacific Ocean, on the 1st May?

With Boston in North America, 1st May?

How do you find what places within the torrid and temperate zones have twelve o'clock at night when it is twelve o'clock at noon at any other place?   35

Which places have twelve o'clock at night when it is twelve o'clock at noon at London?

Which when it is twelve o'clock at noon at Dundee, in Scotland?

Which when it is twelve o'clock at noon at the Gulph of Venice?

Which when it is twelve o'clock at noon at the Gulph of Bothnea, in the Baltic Sea?

How do you find the above-stated circumstance by the planisphere?   36

How do you find by a terrestrial globe all the places to which the sun is rising and setting, and when it is twelve at noon and twelve at night, at a given hour and day at another place?   37

Show this of London for the 20th of May.

Of Dresden in Germany on the 1st of June.

Of Mecca in Arabia on the 1st of April.

How do you find the altitude of the sun at a given place, on a given day, and at a given hour?   38

Figures of re-
ference to the
Problems.

What is the altitude of the sun at Monsul, Turkey in Asia, at eleven o'clock in the morning with them, on the 1st of March?

What on the 1st of May at twelve o'clock at the Bay of Biscay, coast of France?

What on the 1st of June at one o'clock in the afternoon with the inhabitants of Teneriffe, Canary Isles, in the Atlantic Ocean?

How do you find the angle of position between places on the horizon of a globe?　　　39

What is the angle of position between London and Glasgow?

What between the Gulph of Finland and that of Venice?

What between the Gulph of Ormus in the Indian Ocean, and that of Persia in the same ocean?

How do you rectify the celestial globe for the sun's place on a given day, the latitude of a given place, and twelve o'clock at noon at that place?　　　40

Rectify the celestial globe for the ruins of Carthage in Barbary, in Africa, on the 1st of July.

For the island of Oporto in Portugal, 1st of January.

For Olympia in Greece, Turkey in Europe, for the 21st of June.

On the 24th of June for Palermo in Sicily, an island in the Mediterranean, separated from Italy by the straits of Messina.

For Pekin in China, on the 1st of September.

Figures of re-
ference to the
Problems.

What is the position of a right sphere?   Name the circumstances attending that position.   41
    What is the position of the inhabitants of the equator?

What is the position of a parallel sphere?   Name the
circumstances attending that position.   42
    What is the position of the poles?

What is the position of an oblique sphere?   Name the
circumstances attending that position.   43
    What is the position of the inhabitants of London?
    What those of the Canary Isles?
    What of Tyre in Palestine, in Turkey?
    What of the ruins of Troy, Turkey in Asia?

*How do you find the part of the earth that sees an eclipse
of the sun, by a terrestrial globe?*   44

*How do you find by a terrestrial globe the parts of our
earth to which an eclipse of the moon will be visible?*   45

*How do you find by a celestial globe in what part of the hea-
vens a predicted eclipse of the moon will be first seen at London?*   46

*How do you find the sun's meridional altitude for a given
day?*   47
    What is his meridional altitude at London on the 21st
of June?
    What at Seville in Spain on that day?
    What at Spa in Germany on that day?

Figures of re-
ference to the
Problems.

*How do you find the sun's altitude for a given hour on
a given day at a given latitude, and his bearing on the horizon?*   48

What altitude has the sun, and what is his bearing at
the isthmus of Suez in Egypt, Africa, on the 1st of May
at ten in the morning?

What at Upsal in Sweden on that day and hour?

What at the Peak of Teneriffe on that day and hour?

*How do you find the altitude and bearing of a given star
on a celestial globe at a given latitude on a given day and
hour?*   49

What are the altitudes and bearings of the two stars
nearest the head of Ursa Major, called the Pointers, at
London on the 1st of July, at twelve o'clock at noon?

What of Rastaban in the head of Draco, at the same
place, day and hour?

How do you find the latitude and longitude of the stars?   50

What are the latitude and longitude of Alderamin in
Cepheus?

What of Wega in Lyra?

How do you find the latitude and longitude of a planet?   51

*How do you find if a certain planet will be above the ho-
rizon on a given night at a given latitude?*   52

*How do you find the time of the moon being above the ho-
rizon on a given day at a given latitude?*   53

Figures of re-
ference to the
Problems.

How do you find the declination of a fixed star?   54

What is the declination of Deneb Kaitos in Cetus?

What of Arctures in Bootes?

What of Capella?

How do you find the declination of the sun, moon, or a
planet, on a given day?   55

What is the sun's declination on the 1st of November?

What on the 1st of May?

What is meant by the sun's almacanther?   56

What is his almacanther at eleven o'clock in the morning at London on the 1st of August?

What at the same place on the 1st of September at
four o'clock in the afternoon?

How do you find the sun's amplitude at rising and setting
at a given place on a given day?   57

What is his amplitude at rising and setting at Toledo in
Spain on the 1st of April?

What at Waterford in Ireland on the 1st of December?

What at the Royal Observatory, Greenwich, on the
1st of June?

On which two days in the year does the sun rise due east
and set due west to the inhabitants within the torrid and tem-
perate zones?   58

How do you prove that the sun rises due east and sets due
west only twice a year to the inhabitants of an oblique sphere?   59

Figures of re-
ference to the
Problems.

How do you prove that the inhabitants of the equator have always twelve hours day and twelve hours night?   60

Where is the sun's place for the longest day to all places having northern latitude?   61

What is the length of the longest day at London?

Where is the sun's place for the longest day to all places in southern latitude within the torrid and temperate zones?

Where is the sun's place on the shortest day to all places in the above-mentioned latitude? What is the length of the shortest day at London?   62 and 63

What is the length of the shortest day at the Cape of Good Hope?

How do you find by the length of the longest day at a given place what climate it is in?   64

In what climate is the Black Sea in Asia situated?

In what the Caspian Sea in Tartary in Asia?

In what Cape Clear in the Irish Sea?

When the length of the longest day at a place is given, how do you find by it the latitude of the place?   65

What is the latitude of London?

What of Cape Horn, South America?

What of Cape Florida, North America?

What of Corinth in European Turkey?

Resolve by calculation the sixty-fourth question.   66

Figures of re-
ference to the
Problems.

In what climate is Damascus in Turkey in Asia?

In what is the Irish Sea?

*How do you find the points of the compass at which the
sun rises and sets at a given latitude on a given day?*   67

To which points of the compass does the sun rise and
set at London on the 1st of August?

To which on the 3d of May at that place?

To which at the entrance of the Straits of Gibraltar,
between Europe and Africa, on the 4th of June?

To which at the entrance of the Straits of Dover, between England and France, on that day?

To which at the entrance of the Straits of Babelmandel,
between Africa and Asia, in the Red Sea, on the 30th of
April?

To which at the entrance of the Straits of Ormus, between Persia and Arabia, on the same day?

*How do you find what hours on a given day at a given
latitude the sun will be due east and due west*?*   68

What hours on the 19th of June will the sun be due
east and due west to the centre of the Straits of Malacca in
the Indian ocean?

What hours on the 19th of June to the centre of the
Straits of Magellan, South America?

* For the daytime at places in northern latitude, it must be some day when the sun
is north of the equator; for those in southern latitude, when he is south of that circle.

Figures of re-
ference to the
Problems.

*How do you find the day of the month at a given latitude
by the sun's meridional altitude?*   69

What will be the day of the month when the sun's meridional altitude at London is 60°?

What at Edinburgh when the sun's meridional altitude
is 58°?

*How do you find the sun's altitude at any hour of the day
at a given latitude?*   70

What is the sun's altitude at London on the 25th of
June, at four o'clock post meridian?

What at Yarmouth on that day at that hour at that
place?

*Knowing the sun's altitude on a given day at a given la-
titude and at a certain time, how do you find the hour of the
day from it?*   71

What is the hour of the day at London on the 1st of
June, when the sun is 20° above the eastern edge of the
horizon?

What at the centre of the sea of Marmora in Asia, on
the 1st of June, when the sun is 30° above the eastern edge
of the horizon of that latitude?

*How do you find the duration of twilight at a given place
on a given day?*

What is the duration of twilight at the Isle of Lundy
in the Bristol Channel on the 1st of May?

Figures of re-
ference to the
Problems.

What at London on the 1st of May?

*How do you find the duration of twilight at the poles when
the sun has left either of them?*　　　　　　　　73

*How do you find the stars that never rise and those that
never set at a given latitude?　What portion of the celestial
sphere is never seen at the latitude of London?*　　74 and 75

What portion of it is always above the horizon of
London?

What portion of it is always above the horizon of Calcutta, East Indies?

*How do you find if a particular star of those which do
rise and set at a given latitude, will be above the horizon of
that place on a given night?*　　　　　　　　76

Will Spica in Virgo be above the horizon of London
after sun-set on the 1st of October?

Will Autares in Scorpio be above the horizon of Greenwich on the 1st of November after sun-set?

*How do you find the right ascension of a heavenly body at
a given latitude on a given day?*　　　　　　　77

What is the sun's right ascension at London on the 1st
of April?

What on the 15th of July at London?

*What is the right ascension of the sun to the inhabitants
of the equator on the 1st of June?*　　　　　　78

Figures of re-
ference to the
Problems.

What is meant by the oblique ascension and descension of the sun or a star? 79

What is the oblique ascension of the sun at London on the 1st of May?

What the oblique ascension of the bright star in the cheek of Castor on that day?

How do you find the ascensional difference? 80

What is the ascensional difference in the sun on the 1st of June at London?

What is the meaning of azimuth? 81

What is the sun's azimuth at rising on the 1st of May at London?

What on that day at Paris?

How do you find the sun's azimuth and bearing at a given hour of a given day at a given latitude? 82

Find the sun's azimuth and bearing at London on the 1st of May at ten o'clock ante-meridian?

And at Egypt on that day and hour?

How do you find the places of the moon's nodes on a globe? 83

How do you describe the moon's orbit on a globe or armillary sphere? 84

3 B

*How do you find the moon's latitude and longitude, and
the time of her rising and setting at a given latitude on a
given day?*

*How do you find the declination of a given planet on a
given day?*

*Where is the place of the moon when she has the greatest
meridional altitude at London?*
What is her altitude when at 5° 15′ north of the
ecliptic?

*Where is the place of the sun when he has the greatest me-
ridional altitude at London?*
What is his altitude at London when at 23° 30′ of
northern declination, or in the tropic of Cancer?

*For what place of the moon do you rectify the globe for
her least meridional altitude at London?*
What is her meridional altitude at London when she is
5° 15′ south of the equator?

*What is the place of the sun when he has the least meri-
dional altitude at London?*
What is the greatest altitude of the sun at London when
he is in the tropic of Capricorn?

How do you find the times of Regulus in Leo, rising,

Figures of re-
ference to the
Problems.

setting and culminating, or passing the meridian of London
on a given day?   91

When does Procyon in Canis Minor rise, culminate and
set at London on the 1st of January?

How do you find all the stars that are either rising, set-
ting, or on the meridian, and all those above the horizon of a
certain latitude, on a given night and hour?   92

How do you find the day in the year that will be of the
same length of a given day at a given latitude?   93

What day will be the same length at Greenwich with
the 10th of May?

How do you find the hour of the night at a given latitude
by the height of a star?   94

What is the hour of the night at London on the 1st of
December when the middle star in Orion's belt is 30° above
the eastern edge of the horizon of that place?

How do you find the hour of the day by the sun's azimuth?   95

How do you find the latitude of a place by the sun's me-
ridional altitude on a given day?   96

How do you find the place of the sun when he is below
your horizon on a given night?   97

How do you find the relative situation of the sun and

Figures of re-
ference to the
Problems.

*moon to each other when one is above and the other below the
horizon?*　　　　　　　　　　　　　　　　　　　　　　98

Where is the place of the sun when the moon is at full
on the meridian of a given place at twelve o'clock at night?

*How do you find when a star rises with the sun, or cos-
mically, at a given place?*　　　　　　　　　　　　　99

What day does Sirius in Canis Major rise cosmically at
London?

*How do you find when a given star sets cosmically, or at
sun-rise, at a given latitude?*　　　　　　　　　　　　100

When does Fomalhant in the Southern Fish set cosmically?

*How do you find when a star rises and sets achronically,
or at sun-set, at a given latitude?*　　　　　　　　　101

When does Regulus in Leo rise at sun-set, or achronically, at London?

When does the same star set achronically, or at sun-set,
at London?

*How do you find when a star rises and sets heliacally at a
given latitude?*　　　　　　　　　　　　　　　102

When does Aldebaran in Taurus rise heliacally?

When does the same star set heliacally?

*How do you find by a celestial globe if Jupiter or Venus
is a morning star on a given day at a given latitude?*　103

Figures of re-
ference to the
Problems.

*How do you find if Jupiter or Venus is an evening star on
a given day at a given latitude?*   104

*How do you find the hour of the night by the altitude and
azimuth of a star?*   105

*How do you find by two stars having the same azimuth
the hour of the night?*   106

*How do you find by the altitudes of two stars that have
the same azimuth at a given hour the latitude of a place?*   107

*How do you find what night in the year a given star will
be on your meridian at twelve o'clock at night?*   108
When will Rastaban in Draco be on the meridian of
London at twelve o'clock at night?
When will Markab in Cetus?
When will Wega in Lyra?

*How do you find at what hour a given star will be on your
meridian on a given day?*   109
At what hour June 1st will Denebola in Leo be on
the meridian of London?
At what hour June 1st will Alphard in Hydra be on
the meridian of London?
At what hour will Rigel in Orion on that day be on the
meridian of London?

VOCABULARY.

Acute. Sharp, shrill.

Adhesion. Sticking to.

Adjutage. Part of a fountain, being a tube fitted to the aperture of the vessel through which the water is to be thrown.

Aerial. Belonging to the air, as consisting of it, or inhabiting the air.

Aeronaut. A person sailing through the air.

Alkali. Any substance which when mixed with acid produces fermentation.

Altitude. In reference to astronomical observation means the height of the sun, moon, or stars, above the horizon of any place reckoned on a vertical circle from the horizon to the zenith.

Amalgamate. To unite metals with quicksilver.

Ambient. Encompassing.

Analogy. Resemblance between things.

Analyze. To separate the parts of a compound.

Angle. The opening between lines which touch each other in one point.

Aqueous. Watery.

Area. The surface contained between any boundaries.

Armillary. Resembling bracelets, like the circles of the armillary sphere.

Artery. A conical canal conveying the blood from the heart to all parts of the body.

Articulated. Jointed.

Atmosphere. A collection of vapours surrounding the earth.

Attraction. The effect of an invisible natural agent which causes bodies to approach each other.

Attrition. Rubbing.

Auricles. Two appendages of the heart, being two muscular caps containing two ventricles.

Aurora. The morning twilight, or the light occasioned by the refractive power of the atmosphere, which causes the light of the sun to be seen when that luminary is 18° below our horizon.

Axis of a sphere and other solids. A straight line passing through the centre of a body to two opposite points in the circumference. It is that line round which any body will revolve, being the line passing through its centre of gravity. When this line is not spoken of in reference to motion, it is called the diameter of a sphere and the centre of gravity of any other solid body.

Azimuths. Great circles passing through the zenith and perpendicular to the horizon.

Bisected. Divided into two equal parts.

Capillary. Very small like hairs.

Cardinal Points. The east, west, north and south points of the compass. Of the ecliptic, the first degrees of Aries, Cancer, Libra and Capricorn.

Cartilage. A smooth and solid animal substance, softer than bone.

Catoptrics. That part of optics which treats of vision by reflection.

Cellular. Consisting of little cells.

Centrifugal force. The force from the centre of a revolving body.

Centripetal force. That by which a revolving body endeavours to approach a centre.

Cetaceous. Of the whale kind.

Circumfused. Poured round.

Cloud. A collection of vapours in the air.

Coalesce. To unite.

Coeval. Of the same age.

Cohere. To stick together.

Collapse. To contract, to fall together.

Column. A round pillar. Any body pressing vertically on its base.

Combustion. Burning, or consumption by fire.

Commensurate. Having some common measure.

Component. That which constitutes the compound body.

Concentric. Having one common centre.

Concrete. United in one mass.

Concussion. Shock, or shaking.

Condensed. Rendered more close.

Conduit. A canal of pipes for the conveyance of water.

Cone. A figure which has a circle for its base, and terminates in a point.

Continent. A main land which is not interrupted by seas, and so called in contradistinction to island and peninsula.

Converging. Tending to one point from different places.

Corporeal. Having a body.

Corpuscles. Small bodies or atoms.

Cotemporary. Living at the same time.

Cube. A number in arithmetic arising from the multiplication of a square number by its root. Suppose any number, as 3, to be multiplied by itself; the product will be 9, which makes the square. Then if that square number be multiplied by its root, 3, that will make 27; which latter is called the cube, or third power, of 3; and with respect to 27 the number 3 is called the cube root.

Cube. In geometry, a regular body consisting of six square and equal sides.

Day, artificial or solar. The time the sun is above the horizon of any place.

Day, astronomical. That computation of time which begins when the sun transits the meridian of any place and ends at his return to it again.

Day, natural or civil. The time in which the earth completes its revolution on its axis.

Declination. The distance of the sun or a star from the equator north and south. The least declination of a star is its shortest distance from the equator.

Decomposing. Dividing bodies into their component parts.

Degree. The 360 part of a circle.

Denizen. A free man.

Dense. Close, compact.

Desideratum. Something desired, or to be yet discovered.

Develop. To disengage something that is enfolded or concealed.

Diagram. A geometrical representation of any thing.

Diffuse. To spread.

Digit. A measure by which the part of a luminary eclipsed is ascertained. The body eclipsed being always supposed to be divided into twelve parts, called digits, as many of those parts as are eclipsed, so many digits is the body said to be eclipsed.

Direct. In optics means the immediate effect or direction of the primary rays, in contradistinction to those of reflected or refracted ones.

Disc. An apparently plane round surface.

Disseminated. Scattered every way.

Diurnal. Relating to a day, or twenty-four hours.

Divergence. The spreading out of light from a centre.

Ebullition. Boiling or violent motion.

Eclipse. An obscuration of the luminaries of heaven.

Ecliptic circle. The sun's apparent path in the heavens. It is called ecliptic because eclipses happen when the moon is in this circle.

Effervescence. Production of heat by intestine motion.

Efficient. Causing effects.

Effluvia. Small particles flying off from bodies.

Efflux. The act of flowing out.

Elastic. The power of extending and regaining its former shape.

Elementary. The first and constituent principles of any thing.

Emerging. Rising from obscurity.

Ephemeris. An account of the daily motions and situations of the planets.

Epocha. The time at which a new computation is begun. The time from which dates are numbered.

Equator. The circle which equates or divides the globe of our earth into two equal hemispheres, being equally distant from the north and south poles. When we speak of this circle, in reference to the sun or stars, we call it the equinoctial.

Equilateral. Having all sides equal.

Equilibrium. Equality of weight or power.

Equinoctial points, or equinoxes. Those two points of the globe where the ecliptic and equatorial circles intersect each other.

Ether. An element considered to be more subtile than air or light, and an intermediate agent which is supposed to fill the regions beyond the atmosphere of the planets.

Etherial. Subtile, light, celestial.

Evanescent. Vanishing.

Evaporate. To fly off.

Exhaustion. The act of drawing out.

Fallacious. Deceitful.

Fermentation. A slow motion of the intestine particles of a mixed body.

Ferruginous. Denotes a thing to partake of the nature of iron, or to contain particles of that metal.

Finite. Limited, bounded.

Fluid. Any thing that flows easily.

Foci. The plural of focus.

Focus. Where rays meet.

Fossil. Dug out of the bowels of the earth.

Frigid. Cold.

Fulcrum. The prop.

Fusion. The state of being melted.

Galaxy. The milky way, a bright path in the heavens, occasioned by light emitted from an immense number of small stars.

Genus. Species, sort.

Geometrical. Disposed according to the rules of geometry

Geometry. The science of quantity, extension, or magnitude.

Graduated. Marked with degrees.

Gravity. Weight, attraction, tendency to a centre.

Gravity, particular. Is that whereby bodies descend towards the centre of the earth.

Gravity, universal. Is the existence of the same principle in the heavenly regions.

Heliocentric. The place of a planet in which it would appear from the centre of the sun.

Hemisphere. Half a sphere.

Hermetically sealed. Closed with melted glass.

Hesperus. The evening star; an appellation given to Venus when she follows or sets after the sun.

Heterogeneous. Mixed, opposite in nature.

Homogeneous. Having the same nature.

Horizon. The line that terminates our view of the heavens. The rational and the sensible horizon are the same in relation to the fixed stars on account of their immense distance from us: but in regard to the planets they are not so; which produces what is called the horizontal parallax of the sun and planets, when we are considering them.

Humid. Moist.

Humour. Moisture.

Hypothesis. A supposition.

Jet. A spout or shoot of water.

Ignited. Kindled, set on fire.

Impalpable. Not perceivable by the touch.

Impervious. Impenetrable.

Impetus. Force, or violent tendency to any point.

Impinge. To fall on, or strike against.

Impulse. The force of one substance acting on another.

Incidence. The direction in which one body strikes another.

Inclination. The angle made by the orbit of one planet with that of another.

Incorporeal. Without body.

Inert. Inactive, motionless in itself.

Inflammable. Easy to be set on fire.

Inflated. Swelled with the breath.

Inspiration. The act of drawing in the breath.

Insulated. Not contiguous on any side.

Intersection. The point where lines cross each other.

Intuition. Seen by the mind immediately.

Inversely. In a contrary order.

Irradiation. Illumination, emitting beams of light.

Island. A tract of land surrounded by water.

Laboratory. The place where chymical operations are performed.

Latitude. Of places on the earth, is counted from the equator. The latitude of a star is its distance from the ecliptic, north and south, towards the poles of the ecliptic. The latitude of a planet, is its distance from the ecliptic, north and south, within certain boundaries.

Lens. A plane, convex, or concave glass.

Leyden bottle. A glass phial, or jar, coated both inside and out with some conducting substance, for the purpose of being charged.

Ligaments. Strong substances which unite the bones of the body.

Limb. An edge, or border.

Liquefaction. The art of melting; the state of being melted.

Longitude. Of places on the earth, in England is counted from the meridian of London or Greenwich east, and west upon the equator. But longitude of the planets and stars is counted on the ecliptic east and west from the intersection of the ecliptic and equator, at Aries, or the vernal equinoctial point.

Lucid. Bright, glittering.

Lunar aspects. Are those of the moon with the sun and planets; as, opposition, trine, quartile.

Magnet. The loadstone; the stone which attracts iron.

Mathematics. The science which contemplates whatever is capable of being numbered or measured.

Mean distance. That between the greatest and least distance of a planet from the sun.

Mean motion. That between the swiftest and slowest motion of a heavenly body, and being that which would take place if it moved in a perfect circle.

Membrane. A web of several sorts of fibres interwoven for covering some parts of the body.

Mercury. The chemists' name for quicksilver. Also one of the planets.

Meridian. That great circle of the sphere passing over any place and through the north and south poles of the world, being perpendicular to the equator. This is the place of the sun at mid-day, or noon, at every place.

Meridian, magnetical. That to which the poles of the magnet point.

Meteors. Appearances in the sky of a transitory nature, such as clouds, &c.

Mirror. Any thing which represents the images of objects by reflection.

Mode. Form.

Modification. Change in the form, mixture, &c. of any thing.

Momentum. The quantity of force or motion in a moving body.

Muscle. A fibrous fleshy part of the animal body; the immediate instrument of motion.

Nadir. The point immediately under our feet, opposite to our zenith.

Nebulous stars. Those perceived only by a faint light, by which they are not distinctly seen without the aid of glasses; such as those occasioning the appearance called the milky way.

Nerves. The nerves are the organs of sensation, passing from the brain to all parts of the body.

Neutral. Neither acid nor alkaline.

Nocturnal. Nightly.

Nodes. The two points of the ecliptic intersected by the orbit of a planet.

Oblate. Flattened at the poles.

Oblique. Slanting.

Oblong. Longer than broad.

Omnipresent. Present in every place.

Omniscient. Infinitely wise, knowing all things.

Opaque. Not transparent.

Orbit. The line described by the revolution of a planet.

Organization. Construction in which the parts are so disposed as to be subservient to each other.

Oscillation. Moving backwards and forwards like a pendulum.

Particles. Very small parts of matter.

Peninsula. A portion of land joining to a continent by a narrow neck, the rest being encompassed by water.

Penumbra. A faint or partial shade perceived between the perfect shadow and the full light in an eclipse.

Percussion. Striking, pushing.

Phænomenon. (In the plural *Phenomena.*) An extraordinary appearance in nature.

Philosophy. Knowledge natural and moral; the course of the sciences.

Phlogiston. The inflammable part of a body.

Phosphorus. In chemistry, a substance which when exposed to the air takes fire.

Physical. Relating to matter, or natural philosophy. Substance, not spirit.

Piston. A moveable plug in a pump or syringe.

Planet. One of the bodies in our system, which move round and receive light from the sun.

Postulatum. Position assumed without proof.

Precipitated. Thrown to the bottom.

Primary. First.

Problem. A question practically resolved.

Projectile. That force which causes a body to proceed in a straight line.

Promontory. A high land jutting into the sea.

Pulmonary Arteries and Pulmonary Veins. Are vessels which

carry the blood from the heart to the lungs, and those which carry it from the lungs to the heart.

Quadrant. The fourth part of a circle; an astronomical instrument.

Quadrature. The time of the moon entering the first and last quarter.

Quadruple. Fourfold.

Quicksilver. A mineral substance, mercury.

Quiescence. A state of rest.

Radii. The plural of radius.

Radiant. Shining.

Ramify. To separate into branches.

Rarefaction. Lightness or thinness procured by the extension of a body.

Ratio. Proportion.

Rational. Agreeable to reason.

Rationale. A solution or account of the principles of some opinions, action, hypothesis, phenomenon.

Ray of light. The least particle of light that can be separately impelled.

Rays, converging. Those which tend to a point. All convex lenses make the rays converge, and concave ones make them diverge.

Reaction. The resistance any body makes to a force impressed on it. The action and reaction of bodies upon each other are equal.

Receiver. The vessel of the air-pump out of which the air is drawn to receive any body on which experiments are to be tried.

Rectilineal. Right-lined.

Reflection. The act of throwing back.

Refraction. The turning out of the straight course of a body in motion. The variation of a ray of light from the straight line which it would have passed on in, had not the rays passed through a new medium.

Repulsive. Having the power to beat back.

Respiration. The act of breathing.

Retrograde. Being removed backward.

Retrogression. Going backwards.

Satellite. A small planet attendant on a larger one.

Saturated. Fully impregnated.

Science. Knowledge founded on demonstration.

Section. A cut or division which separates one part from the rest.

Sensible. Perceivable by one of the five senses.

Species. Sort, kind, &c.

Specific gravity. Comparative weight bulk for bulk.

Speculum. A looking-glass.

Spheroid. A body, oblate or oblong, approaching to the form of a sphere.

Spiral. Winding, curved.

Square number. The product of a number multiplied by itself, as 4 multiplied by itself, 4 times 4, are 16, which is the square of 4; and 4 is the square root of 16.

Stellated. In the form of stars.

Subdivide. To divide a part into more parts.

Sublunary. Beneath the moon.

Subtend. To extend under certain limitations.

Subtile. Thin, fine, rare.

Superfice. The surface of a solid. Lines are the boundaries of a superfices; a point is the extremity of a line; a plane is a level or flat superfice.

Synonymous. Expressing the same thing in different terms.

System. A scheme which unites many things in order.

Systematic. In a regular way

Tacitly. Silently.

Tangible. Perceptible by the touch.

Tegument. A covering; the outward part.

Terraqueous. Composed of land and water.

Terrestrial. Earthly.

Theory. Not practice.

Torpedo. A fish which when alive if touched even with a stick benumbs the hands, but when dead is eaten safely.

Torpid. Benumbed.

Torrid. Burning. The zone of the earth between the tropics, where the rays of the sun fall perpendicularly.

Transit. The passage of a planet over the face of the sun or another planet, or some line of the heavens.

Transverse. Being in a cross direction.

Trigonometry. The art of measuring triangles.

Vacuum. A space devoid of air or matter.

Vapour. Any effluvium from terrestrial substances that mingles with the air.

Vernal. Belonging to spring.

Vertical. Perpendicular to the horizon.

Vibrate. To move to and fro.

Vital. Essential to life.

Vitiate. To render impure.

Vivid. Lively, striking, active.

Vivifying. Animating.

Volatile. Parts flying off. Airy.

Undulation. A waving motion.

Zenith. The point directly over our heads.

Zodiac. A girdle which surrounds the apparent sphere of the heavens, and extends 8° on each side of the ecliptic, including the orbits of the planets.

Zone. A girdle, or belt.

THE END.

Printed by T. DAVISON,
    Whitefriars.

CORRECTIONS OF LETTERS AND FIGURES OF REFERENCE.

Page 27, line 23, for *AB*, read *AD*.——for Plate 5, read Plate 2.

43, 17, for *B*, read *D*.——Page 43, line 18, for *C*, read *E*.

43, 22, dele *A*, and for *C*, read *E*.——Page 72, line 16, for 6, read 2.

72, 17, dele *B*, and for 2, read 5.——Page 74, line 23, for 4, read 7.

90, 19, for 2, read 6.——Page 92, line 15, dele *A*.

92, 16, dele *B*.——Page 92, line 19, dele *A*.——Page 92, line 20, dele *A*.

92, 23, dele *B*.——Page 130, line 8, for *A*, read *D*.

130, 13, introduce *B* before *fig*. 2.——Page 131, line 16, for *fig*. 1, read *fig*. 4.

132, 5, for *D*, read *A*.——Page 133, line 2, dele *E* and *G*.

134, 11, for 15, read 14.——Page 154, line 10, dele *A B*.

170, 11, for *C*, read *L*.——Page 181, line 5, dele *A*, *fig*. 14.

186, 7, for *A*, read *C*.——Page 190, line 9, for 7, read 5.

190, 10, for 7, read 5.——Page 210, line 10, for 1 : 2, read 2 : 3.

224, 26, for *A*, read *D*.——Page 226, line 24, dele *B*.

236, 23, for 5, read 4.——Page 236, line 24, instead of *n n*, read *n m*.

239, 5, dele *i d* and *e*.——Page 306, line 8, dele *N*.

307, 25, introduce *fig*. 1, plate 3, after *b*.——Page 307, line 26, for *d*, read *o*.

308, 1, for *E*, read *b*.——Plate 30, *fig*. 4, dele *D*, and introduce *A* at the top part of *fig*. 4.

The Authoress not being able regularly to procure the Engravings to compare with the Letter-press, has to apologize for the errors that have consequently arisen in the letters and figures of reference to the plates, but which may be corrected by a pen agreeably to the directions above given.